Effective Maintenance Management

Risk and Reliability Strategies for Optimizing Performance

by

V. Narayan

Industrial Press
New York

Library of Congress Cataloging-in-Publication Data

Narayan, V.

Effective maintenance management : risk and reliability strategies for optimizing performance / V. Narayan.

p. cm

ISBN 0-8311-3178-0

1. Plant maintenance—Management. 2. Reliability (Engineering). 3. Risk Management. I. Title.

TS192.N355 2003

658.2—dc22

2003055881

Effective Maintenance Management

Interior Text and Cover Design: Janet Romano

Managing Editor: John Carleo

Industrial Press Inc.

200 Madison Avenue

New York, New York 10016

10 9 8 7 6 5 4 3 2 1

Dedication

To the memory of my late parents Venkatraman and Meenakshi, and my late mother-in-law Saraswathy.

Foreword

A few years ago, I met the author at an international conference in Exeter. Over lunch, he outlined some of his ideas on risk management and system effectiveness - with a fork and a couple of knives as props. I found his approach refreshing, different and worth pursuing. These were largely in line with my own views on System Operational Success, so I encouraged him to write this book.

As the author had many years of experience in the maintenance of Refineries, Gas plants, Offshore Platforms as well as Engineering and Pharmaceutical plants, I thought that the blend of theory and practice would be useful. The relevance of theory is brought home with a number of illustrative examples from industrial situations, so I feel my point is well made. His approach to maintenance is holistic and as such it could be applied to situations involving financial risk, public health or the maintenance of law and order.

He explains the raison d'être of maintenance; this should help maintenance managers justify their efforts rationally. The discussion on risk perceptions and why they are important may strike a chord with many of us. Knowing what tools to use and where to apply them is important as also how to manage data effectively.

The book will help maintenance managers, planners and supervisors, as well as students in understanding how best to reduce industrial risks. This should help them improve both technical and production integrity, leading to fewer safety, health and environmental incidents while increasing the quality and production levels and reducing costs.

Dr. Jezdimir Knezevic
MIRCE Akademy

Acknowledgements

I have had the good fortune to work with talented and experienced people throughout my working life. In writing this book, I have used the knowledge gained from the interaction with the thinking and work of many colleagues, associates and friends. I hope I have done justice to all of them, and that readers, especially maintenance practitioners, apply the knowledge they gain, by improving the effectiveness of maintenance in their Industry.

I met Dr. Jezdemir Knezevic in 1998 by chance at a business party. In the course of a brief discussion, I realized that I had met a teacher who worked in the real world. His practical and yet mathematically sound approach was refreshing, and a friendship began almost immediately. He, in turn, encouraged me to write this book. Once I completed the manuscript, he reviewed it critically and made a number of useful suggestions to improve both content and readability.

My colleagues at work, both current and past, have helped shape my thinking. They are too numerous to name, so I will not attempt this task. One in particular, Greg Stockholm, of Shell Chemicals Inc., taught me the Root Cause Analysis process. While problem solving and failure analysis are essential skills for every maintainer, I had not realized till after Greg's course that there was a structured and logical process available. I have in turn, trained many others to use Root Cause Analysis and Reliability Centered Maintenance. The feedback and comments made by several hundred students from around the world have helped shape my ideas.

Many friends, some in Academia and others in Industry or Consultancy firms have reviewed the book. Their comments and advice have been invaluable. The reviewers include Paddy Kirrane, Jim Wardaugh, Nick Schulkins, Roger Peres, Rob Knotts, Jane Goodacre and Dr. Helge Sandtorv. At his initiative, Dr. Sandtorv forwarded a copy to Prof. Marvin Rausand of the Norwegian University of Science and Technology. Dr. Rausand graciously reviewed the first six chapters, and provided detailed comments. My son Sharad, daughter Shweta, father-in-law Ramamurthy and a family friend Rajesh, reviewed it from the lay persons' point of view. All the reviewers have been very constructive and assisted me greatly in bringing the book to print.

I offer my thanks to Shell U.K Exploration and Production, who allowed me to proceed with this effort while I was working with them.

Keith and Valerie Pottinger converted my sketches and charts into presentable material, corrected my grammar, paginated and brought this book to its finished form. Notwithstanding many pressures on their time, they have juggled their schedules to help a friend. My friend Richa Saxena used her graphic arts skills to draft several sample book cover designs. My publishers, Industrial Press Inc. have been very helpful from the time I first made contact with them. I thank John Carleo, Janet Romano and Patrick Hansard for their constructive and supportive approach. Janet and John have also helped me out by arranging to incorporate the final set of corrections and prepare the manuscript for printing. Throughout the time I was busy with the book my wife Lata has been patient and considerate, thoughtful and helpful. I am sure I could not have made it without her love and support.

To all of them, I offer my gratitude.

Preface

The traditional view of the general public towards maintenance is one of elegant simplicity. Their contact is often limited to automobile or appliance repair workshops. From this experience, maintenance appears to be an unavoidable activity that costs money and takes time. The view held in the board rooms of industry appears to match this perception.

Good news is generally not news at all, so people only tend to think of maintenance when things go badly wrong. The moment there is a major safety or environmental incident, the media come alive with news of the maintenance cutbacks, real or imaginary, that have allegedly contributed to the incident. Think of what you saw on TV or read in the newspapers after any of the airline, ferry or industrial disasters, and you will readily recognize this picture.

What do we actually do when we manage a business? In our view, we manage the risk—of safety and environmental incidents, adverse publicity, loss of efficiency or productivity, and loss of market share. A half century ago, Peter F. Drucker[1], a well known management guru, said:

"It is an absolute necessity for a business enterprise to produce the profit required to cover its future risks, to enable it to stay in business and to maintain intact its wealth producing capacity."

This is as valid today as it was then. In the maintenance management context, the risks that are of concern to us relate to safety or environmental incidents, adverse publicity, and of loss of profitability or asset value.

We will examine the role of maintenance in minimizing these risks. The level and type of risks vary over the life of the business. Some risk reduction methods work better than others. The manager must know which ones to use, as the cost-effectiveness of the techniques differ. We will look at some of the risk reduction tools and techniques available to the maintainer, and discuss their applicability and effectiveness.

Risks can be quantitative or qualitative. We can usually find a solution when dealing with quantified risks, which relate to the probability and consequence of events. Qualitative risks are quite complex and more difficult to resolve, as they deal with human perceptions. These relate to peoples' emotions and feelings and are difficult to pre-

1 The Practice of Management, page 38, first published by William Heinemann in 1955. Current edition is published by HarperBusiness, 1993, ISBN 0887306136.

dict or sometimes even understand. Decision-making requires that we evaluate risks, and both aspects are important. The relative importance of the qualitative and quantitative aspects of risk varies from case to case and person to person. Even the same person may use a different recipe each time. We should not categorize people or businesses as risk-seeking or risk-averse. It is not merely a mind-set; the situation they face determines their attitude to risk. All these factors make the study of risk both interesting and challenging.

In this book, we set out to answer three questions:

- *Why do we do maintenance and how can we justify it?*
- *What are the tasks we should do to minimize risks?*
- *When should we do these tasks?*

We have not devoted much time to the actual methods used in doing various maintenance tasks. There are many books dealing with the how-to aspects of subjects such as alignment, bearings, lubrication, or the application of Computerized Maintenance Management Systems. Other books deal with organizational matters or some specific techniques such as Reliability Centered Maintenance. We have concentrated on the risk management aspects and the answers to the above questions.

Throughout this book, we have kept the needs of the maintenance practitioner in mind. It is not necessary for the reader to have knowledge of systems and reliability engineering. We have devoted a chapter to develop these concepts from first principles, using tables and charts in preference to mathematical derivations. We hope that this will assist the reader in following subsequent discussions. Readers who wish to explore specific aspects can refer to the authors and publications listed at the end of each chapter. There is a glossary with definitions of terms used and a list of acronyms and abbreviations at the end of the book.

We believe that maintainers and designers can improve their contribution by using reliability engineering theory and the systems approach, in making their decisions. A large number of theoretical papers are available on this subject, but often they are abstract and difficult to apply. So these will remain learned papers, which practitioners do not understand or use. This is a pity because maintainers and designers can use the help which reliability engineers can provide. We hope that this book will help bridge the chasm between the designers and maintainers on the one hand, and the reliability engineers on the other. In doing so, we can help businesses utilize their assets effectively, safely, and profitably.

Table of Contents

Abbreviations & Acronyms

The following abbreviations and acronyms have been used in the book.

Term	Full Expression	Refer also to
AGAN	As Good As New	
AIDS	Acquired Immune Deficiency Syndrome	
ALARP	As Low As Reasonably Practicable	
BSE	Bovine Spongiform Encephalopathy	
CAPEX	Capital Expenditure	
CJD	Creutzfeldt-Jakob Disease	
CMMS	Computerized Maintenance Management System	
ESD	Emergency Shutdown	
FBD	Functional Block Diagram	IDEF
FCA	Failure Characteristic Analysis	
FMEA	Failure Modes and Effects Analysis	
FPSO	Floating Production, Storage and Offloading vessel	
FTA	Fault Tree Analysis	
HAZOP	Hazard And Operability Study	
HSE	Health and Safety Executive	
ICAM	Integrated Computer-Aided Manufacturing	
IDEF	Icam-DEFinition	ICAM
JIP	Joint Industry Project	
J-T	Joule-Thomson	
KISS	Keep It Simple, Stupid!	
MLE	Maximum Likelihood Estimator	
MTBF	Mean Operating Time Between Failures	MTTF, MTTR
MTTF	Mean Time To Failure	MTBF, MTTR
MTTR	Mean Time To Restore	MTTF,MTBF
NASA	National Aeronautics and Space Administration	

Term	Full Expression	Refer also to
NPSH	Net Positive Suction Head	
OIM	Offshore Installation Manager	
OPEX	Operating Expenditure	
OREDA	Offshore Reliability Data	
pdf	Probability Density Function	
PRV	Pressure Relief Valve	PSV
PSV	Pressure Safety Valve	PRV
PTW	Permit To Work	
RBD	Reliability Block Diagram	
RCA	Root Cause Analysis	
RCM	Reliability Centered Maintenance	FMEA, FCA
SMS	Safety Management System	
TNT	Trinitrotoluene	
TPM	Total Productive Maintenance	

The Production and Distribution Process

CHAPTER 1

This book deals with the management of risks through the life cycle of a process plant. We will address the question of **why** we do maintenance, **what** tasks we actually need to do, and **when** we should do them, so as to reduce these risks to a tolerable level and an acceptable cost. We will examine the role of maintenance in obtaining the desired level of system effectiveness, and begin this chapter with a discussion of the production and distribution process. After going through this chapter, the reader should have a better appreciation of the following:

- The production and distribution process and its role in creating value as goods and services;
- Difficulties in measuring efficiency and costs; understanding why distortions occur;
- Determination of value and sources of error in measuring value;
- Reasons for the rapid growth in both manufacturing and service industries;
- Understanding the systems approach; similarities in the manufacturing and service industries;
- Impact of efficiency on the use of resources;
- Maintenance and the efficient use of resources.

We need goods and services for our existence and comfort; this is, therefore, the focus of our efforts. We change raw materials into products that are more useful. We make, for example, furniture from wood or process data to obtain useful information. By doing so, we add value to the raw materials, thereby creating products that others need. We can also add value without any physical material being used. Thus, when a nurse takes a patient's temperature, this information helps in the diagnosis of the illness, or in monitoring the line of treatment. Another instance of adding value is by bringing a product to the market at the right time. Supermarkets serve their customers by stocking their shelves adequately with food (and other goods). They will not be willing to carry excessive stocks as there will be wastage of perishable goods. Overstocking will also cost the supermarket in terms of working capital, and therefore reduce profit margins. By moving goods to the shelves in time, supermarkets and their customers benefit, so we conclude that their actions have added value. The term *distribution* describes this process of movement of goods. It adds value by increasing consumer access.

Production processes include the extraction of raw materials by mining, and their conversion into useful products by manufacturing. If the main resource used is physical or intellectual energy, with a minimum of raw materials, we call it a service. The word *process* describes the flow of work, which enables production of goods or provision of services. In every commercial or industrial venture there is a flow of work, or Business Process. The business can vary widely; from a firm of accountants to a manufacturer of chemicals to a courier service.

In the case of many service industries, the output is information. Lawyers and financial analysts apply their knowledge, intellect, and specialized experience to process data and advise their clients. Management consultants advise businesses, and travel agents provide itinerary information, tickets, and hotel reservations. In all these cases, the output is information that is of value to the customer.

1.1 PROCESS EFFICIENCY

1.1.1 Criteria for assessing efficiency

In any process, we can obtain the end result in one or more ways. When one method needs less energy or raw materials than another, we say it is more efficient. For a given output of a specified quality, the process that needs the least inputs is the most efficient.The process can be efficient in respect of energy usage, materials usage, human effort, or other selected criteria.Potential damage to the environment is a matter of increasing concern, so this is an additional criterion to consider.

If we try to include all these criteria in defining efficiency, we face some practical difficulties. We can measure the cost of inputs such as materials or labor, but measuring environmental cost is not easy. The agency responsible for producing some of the waste products will not always bear the cost of minimizing their effects. In practical terms, efficiency improvements relate to those elements of cost that we can measure and record. It follows that such incomplete records are the basis of some efficiency improvement claims.

1.1.2 Improving efficiency

Businesses try to become more efficient by technological innovation, business process re-engineering, or restructuring. Efficiency improvements that are achieved by reducing energy inputs can impact both the costs and undesirable by-products. In this case, the visible inputs and the undesirable outputs decrease, so the outcome is an overall gain. A similar situation arises when it comes to reducing the input volume of the raw materials or the level of rejections.

When businesses make efficiency improvements through workforce reductions, complex secondary effects can take place. If the economy is buoyant, there may be no adverse effect, as those laid-off are likely to find work elsewhere. When the economy is not healthy, prevailing high unemployment levels will rise further. This could perhaps result in social problems, such as an increase in crime levels. The fact that workforce reductions may sometimes be essential for the survival of the business complicates this further. There may be social legislation in place preventing job losses, and as a result, the firm itself may go out of business.

1.1.3 Cost measurement and pitfalls

There are some difficulties in identifying the true cost of inputs. What is the cost of an uncut piece of diamond or a barrel of crude oil? The cost of mining the product is all that is visible, so this is what we usually understand as the cost of the item. We can add the cost of restoring the mine or reservoir to its original state, after extracting the ores that are of interest, and recalculate the cost of the item. We do not calculate the cost of replenishing the ore itself, which we consider as free.

Let us turn to the way in which errors can occur in recording costs. With direct or activity based costing, we require the cost of all the inputs. This could be a time-consuming task, and can result in delays in decision making. In order to control costs, we have to make the decisions in time.

Good accounting practice mandates accuracy, and if for this purpose it takes more time, it is a price worth paying. Accounting systems fulfill their role, which is to calculate profits, and determine tax liabilities accurately. However, they take time, making day-to-day management difficult. Overhead accounting systems get around this problem by using a system of allocation of costs. These systems are cheaper and easier to administer. However, any allocation is only valid at the time it is made, and not for all time. The bases of allocation or underlying assumptions change over time, so errors are unavoidable. This distorts the cost picture and incorrect cost allocations are not easy to find or correct.

Subsidies, duty drawbacks, tax rebates, and other incentives introduce other distortions. The effect of these adjustments is to reduce the visible capital and revenue expenditures, making an otherwise inefficient industry viable. From an overall economic and political perspective, this may be acceptable or even desirable. It can help distribute business activity more evenly and relieve overcrowding and strain on public services. However, it can distort the cost picture considerably and prevent the application of market forces.

We have to recognize these sources of errors in measuring costs. In this book we will use the concept of cost as we measure it currently, knowing that there can be some distortions.

1.2 WORK AND ITS VALUE

1.2.1 Mechanization and productivity

When we carry out some part of the production or distribution process, we are adding value by creating something that people want. We have to measure this value first if we want to maximize it. Let us examine some of the relevant issues.

In the days before the steam engine, we used human or animal power to carry out work. The steam engine brought additional machine power, enabling one person to do the work that previously required several people. As a result each worker's output rose dramatically. The value of a worker's contribution, as measured by the number of items or widgets produced per hour, grew significantly. The wages and bonuses of the workers kept pace with these productivity gains.

1.2.2 Value added and its measurement

We use the cost of inputs as a measure of the value added, but this approach has some shortcomings. Consider 'wages' as one example of the inputs. We have to include the wages of the people who produced the widgets, and that of the truck driver who brought them to the shop. Next we include the wages of the attendant who stored them, the salesperson who sold them, and the store manager who supervised all this activity. Some of the inputs can be common to several products, adding further complexity. For example, the store manager's contribution is common to all the products sold; it is not practical to measure the element of these costs chargeable to the widgets under consideration. We have to distribute the store manager's wages equitably among the various products, but such a system is not readily available. This example illustrates the difficulty in identifying the contribution of wages to the cost. Similarly, it is difficult to apportion the cost of other inputs such as heating, lighting, or ventilation.

We can also consider 'value' from the point of view of the customers. First, observe the competition, and see what they are able to do. If they can produce comparable goods or services at a lower price than we can, customers will switch their loyalty. From their point of view, the value is what they are willing to pay. The question is: how much of their own work are they willing to barter for the work we put into making the widgets? Pure competition will drive producers to find ways to improve their efficiency, and drive prices downwards. Thus, another way is to look at the share of the market we are able to corner. Using this approach, one could say that Company A, which commands a larger share of the market than Company B, adds more value. Some lawyers, doctors, and consultants command a high fee rate because the customer perceives their service to be of greater value.

Assigning a value to work is not a simple task of adding up prices or costs. We must recognize that there will be simplifications in any method used, and that we have to make some adjustments to compensate for them. Efficiency improvements justified on cost savings need careful checking—are the underlying assumptions and simplifications acceptable?

1.3 MANUFACTURING AND SERVICE INDUSTRIES

1.3.1 Conversion processes

We have defined manufacturing as the process of converting raw materials into useful products. Conversion processes can take various forms. For example, an automobile manufacturer uses mainly physical processes, while a pharmaceutical manufacturer primarily uses chemical or biological processes. Power generation companies that use fossil fuel use a chemical process of combustion and a physical process of conversion of mechanical energy into electrical energy. Manufacturers add value, using appropriate conversion processes.

1.3.2 Factors influencing the efficiency of industries

Since the invention of the steam engine, the productivity of human labor has increased steadily. Some of the efficiency gains are due to improvements in the production process itself. Inventions, discoveries, and philosophies have helped the process. For example, modern power generation plants use a combined-cycle process. These use gas turbines to drive alternators. The hot exhaust gases from the gas turbines help raise high-pressure steam that provides energy to steam turbines. These drive other alternators to generate additional electrical power. Thus, we can recover a large part of the waste heat, thereby reducing the consumption of fuel.

A very significant improvement in productivity has occurred in the last quarter of the twentieth century due to the widespread use of computers. With the use of computers, the required information is readily available, thereby improving the quality and timeliness of decisions.

1.3.3 Factors affecting demand

The demand for services has grown rapidly since the second World War. Due to the rise in living standards of a growing population, the number of people who can afford services has grown dramatically. As a result of the larger demand and the effects of economies of scale, unit prices have kept falling. These, in turn, stimulate demand, accounting for rapid growth of the services sector. In the case of the manufacturing sector, however, better, longer lasting goods have reduced demand somewhat.

Demographic shifts have also taken place, and in many countries there is a large aging population. This has increased the demand for health care, creating a wide range of new service industries. Similarly, concern for the environment has led to the creation and rapid growth of the recycling industry.

1.4 THE SYSTEMS APPROACH

Some of the characteristics of the manufacturing and service industries are very similar. This is true whether the process is one of production or distribution.We will consider a few examples to illustrate these similarities.

A machinist producing a part on an automatic lathe has to meet certain quality standards, such as dimensional accuracy and surface finish. During the machining operation, the tool tip will lose its sharpness. The machine itself will wear out slightly, and some of its internal components will go out of alignment. The result will be that each new part is slightly different in dimensions and finish from the previous one. The parts are acceptable as long as the dimensions and finish fall within a tolerance band. However, the part produced will eventually fall outside this band. At this point, the process has gone out of control, so we need corrective action. The machinist will have to replace the tool and reset the machine, to bring the process back in control. This is illustrated in Figure 1.1.

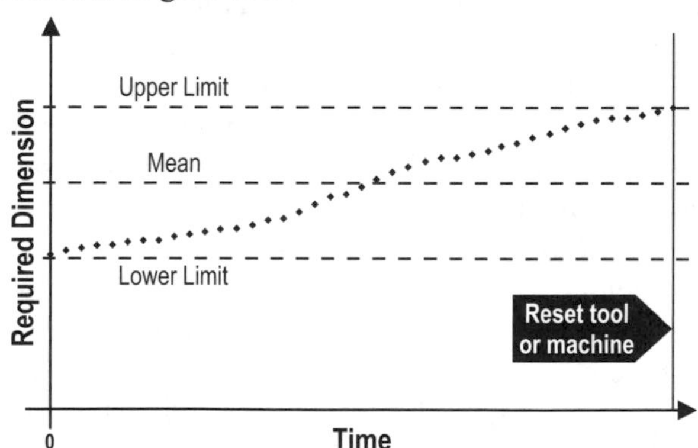

Figure 1.1 Process control chart.

In a chemical process plant, we use control systems to adjust the flow, pressure, temperature, or level of the fluids. Consider a level-controller on a vessel. The level is held constant, within a tolerance band, using this controller. Referring to Figure 1.2, the valve will open more if the level reaches the upper control setting, allowing a larger outward flow. It will close to reduce flow, when the liquid reaches the lower control setting. As in the earlier example, here the level-controller helps keep the process in control by adjusting the valve position.

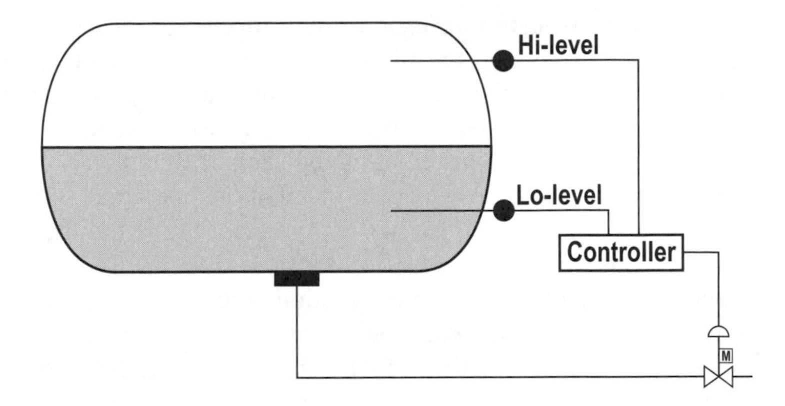

Figure 1.2 Level controller operation.

Consider now a supermarket that has a policy of ensuring that customers do not have to wait for more than 5 minutes to reach the check-out counter. Only a few check-out counters will be open during slack periods. Whenever the queues get too long, the manager will open additional check-out counters. This is similar to the control action in the earlier examples.

Companies use internal audits to check that staff observe the controls set out in their policies and procedures. Let us say that invoice processing periods are being audited. The auditor will look for deviations from norms set for this purpose. If the majority of the invoices take longer to process than expected, the process is not in control. A root cause analysis of the problem will help identify reasons for the delays.

Though these examples are from different fields of activity, they are similar when seen from the systems point of view. In each of these examples, we can define the work flow by a process, which is subject to drift or deviation. If such a drift takes place, we can see it when the measured value falls outside the tolerance band. The process control mechanism then takes over to correct it. Such a model allows us to draw generalized conclusions that we can apply in a variety of situations.

1.5 IMPACT OF EFFICIENCY ON RESOURCES

1.5.1 Efficiency of utilization

Earlier, we looked at some of the factors influencing the efficiency in the manufacturing phase. For this purpose, we define efficiency as the ratio of the outputs to the inputs. We can also examine the way the consumer uses the item. We define efficiency of utilization as the ratio of the age at which we replace an item to its design life under the prevailing operating conditions.

First, we examine whether we use the item to the end of its economic life. Second, is it able to reach the end of its economic life? In other words, do we operate and maintain it correctly? If not, this can be due to premature replacement of parts. When we carry out maintenance on a fixed time basis, useful life may be left in some of the parts replaced. Alternatively, we may replace parts prematurely because of poor installation, operation, or maintenance. In this case, the part does not have any useful life left at the time of replacement, but this shortening of its life was avoidable.

Manufacturers are concerned with production efficiency because it affects their income and profitability. From their point of view, if the consumer is inefficient in using the products, this is fine, as it improves the demand rate for their products. Poor operation and maintenance increases the consumers' costs. If these consumers are themselves manufacturers of other products, high operating costs will make their own products less profitable. *This book helps the consumer develop strategies to improve the efficiency of utilization.*

1.5.2 Efficiency and non-renewable resources

An increase in efficiency, whether it is at the production or at the consumption end, reduces the total inputs and hence the demand for resources. We can ease the pressure on non-renewable resources greatly by doing things efficiently. In this context, the efficiency of both producer and consumer are important.

The first step in improving efficiency is to measure current performance. Qualitative or subjective measurements are perfectly acceptable and appropriate in cases where quantitative methods are impractical.

1.6 MAINTENANCE—THE QUESTIONS TO ADDRESS

When an item of equipment fails prematurely, we incur additional maintenance costs and a loss of production. As a result we cannot utilize the full capability of the equipment. Timely and effective maintenance helps avoid this situation. Good maintenance results in increased production and reduced costs. Correct maintenance increases the life of the plant by preventing premature failures. Such failures lead to inefficiency of utilization and waste of resources. This explains **why** we need to maintain equipment and we will examine it further in chapter 9. There, we will see the essential role of maintenance is to ensure the viability and profitability of the plant. In chapter 10, we offer guidance on the strategies available to you to find the most applicable and effective tasks and to select from these the ones with the lowest cost. At the end of chapter 10, you should have a clear idea of **what** tasks are required and **when** they should be done in order to manage the risks to viability and profitability of the plant.

1.7 CHAPTER SUMMARY

We began this chapter by defining the production and distribution process and then looked at some of the factors that influence efficiency. We use costs to measure performance; low costs imply high efficiency. When measuring costs, we make simplifications, as a result of which we may introduce some distortions.

We discussed how we compute the value of work, using production costs or competitive market prices. We noted that there are some sources of error in arriving at the value of work.

Thereafter, we saw how manufacturing and service industries add value. Manufacturing productivity has grown dramatically, due to cheap and plentiful electro-mechanical power and, more recently, computing power. A beneficial cycle of increased productivity, raising the buying power of consumers, results in increased demand. This has lowered prices further, encouraging rapid growth of manufacturing and services industries.

Manufacturing and service industries are similar processes. The systems approach helps us to understand these, and how they to control them. We illustrated this similarity with a number of examples.

Thereafter, we examined the impact of efficiency on the use of resources. We note that cost is a measure of efficiency, but recognize that all costs are not visible; hence distortions can occur. With this understanding, we saw how to use costs to monitor efficiency. We will address the questions **why, what** and **when** in regard to maintenance as we go through the book.

CHAPTER 2

Process Functions

The term *process* describes the flow of materials and information. In order to achieve our business objectives, we use energy and knowledge to carry out the process.

The purpose of running a business is to produce or distribute goods (or services) efficiently. A business uses its mission statement to explain its objectives to its customers and staff. This is a top-down approach and enables us to see how to fulfill the mission, and what may cause mission-failure. We call this a functional approach, because it explains the purpose, or function, of the business. We can judge the success or failure of the business by seeing if it has fulfilled its function, as described in the mission statement. A high-level function can be broken down into sub-functions. These, in turn, can be dissected further, all the while retaining their relationship to the high-level function.

After reading the chapter, readers who are unfamiliar with this approach should have acquired an understanding of the method—this is the mission or function of this chapter. The main elements of the method are as follows:

- The functional approach, methodology, and communication;
- Identification of functional failure, use of Failure Modes and Effects Analysis, and consequence of failures;
- Reduction of frequency and mitigation of the consequences of failures;
- Cost of reducing risks;
- Damage limitation and its value.

2.1 THE FUNCTIONAL APPROACH

The U.S. Air Force initiated a program called Integrated Computer Aided Manufacturing (ICAM) in the 1970s. They developed a simple tool to communicate this program to technical and non-technical staff, named ICAM-DEFinition or IDEF methodology[1, 2]. With IDEF, we use a graphical representation of a system using activity boxes to show what is expected of the system. Lines leading to and from these boxes show the inputs, outputs, controls, and equipment.

As an illustration, consider a simple pencil. What do we expect from it? Let us use a few sentences to describe our expectations.

A. To be able to draw lines on plain paper.
B. To be able to renew the writing tip when it gets worn.
C. To be able to hold it in your hand comfortably while writing.
D. To be able to erase its markings with a suitable device (eraser).
E. To be light and portable, and to fit in your shirt pocket.

The item must fulfill these functional requirements or you, the customer, will not be satisfied. If any of the requirements are not met, it has failed. Figure 2.1 illustrates a functional block diagram (FBD) of how we represent the second function in a block diagram.

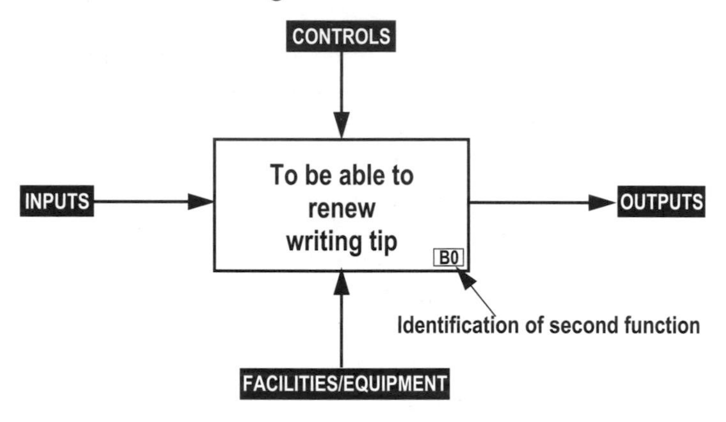

Figure 2.1 FBD of pen system.

Note that we state our requirements in the most general way possible. Thus it does not have to be a graphite core held in a wooden stock. It can easily be a metal tube holder, and still meet our requirements. The second function is met whether we have a retractable core or if we have to shave the wood around the core. It could have a hexagonal or circular section, but must be comfortable to hold. The writing medium cannot be ink, as it has to be erasable. Finally, its dimensions and weight are limited by the need for comfort and size of your shirt pocket!

Every production or distribution process has several systems, each with its own function, as illustrated by the following examples.

- A steam power-generation plant has a steam-raising system, a power generation system, a water treatment system, a cooling system, a control and monitoring system, and a fire protection system.
- A courier service has a collection and delivery system, storage and handling system, transport system, recording and tracking system, and an invoicing system.
- An offshore oil and gas production platform has a hydrocarbon production system, an export system, a power generation system, a communication

system, a fire and gas protection system, a relief and blow-down system, an emergency shutdown system, and a personnel evacuation system.
- A pizza business with a home delivery service has a purchasing system, a food preparation system, a communication system, and a delivery system. Sometimes, all these systems may involve just one person, who is the owner-cook-buyer-delivery agent!

We can use functional descriptions at any level in an organization. For example, we can define the function of a single item of equipment. Jones[3] illustrates how this works, using the example of a bicycle, which has the following sub-systems:

- Support structure, e.g., the seat and frame;
- Power transmission, e.g., pedals, sprockets, and drive chain;
- Traction, e.g., wheels and tires;
- Steering, e.g., handles and steering column;
- Braking, e.g., brakes, brake levers, and cables;
- Lighting, e.g., dynamo, front and back lights, and cables.

We can define the function of each sub-system. For example, the power transmission system has the following functions:

- Transfer forces applied by rider to drive-sprocket;
- Apply forces on chain;
- Transmit the force to driven-sprocket to produce torque on rear wheel.

Similarly, we can examine the other sub-systems and define their functions. The functional failure is then easy to define, being the opposite of the function description; in this case, fails to transfer force.

2.2 FUNCTIONAL BLOCK DIAGRAMS (FBD)

These systems and sub-systems below them are aligned to meet the overall objectives. An FBD is an effective way to demonstrate how this works. It illustrates the relationship between the main function and those of the supporting systems or sub-systems.

We describe the functions in each of the rectangular blocks. On the left side are the inputs—raw materials, energy and utilities, or services. On the top we have the systems, mechanisms, or regulations that control the process. The outputs, such as intermediate (or finished) products or signals, are on the right of the block. Below each block, we can see the means used to achieve the function; for example, the hardware or facilities used to do the work. *As a result of this approach, we move away from the traditional focus on equipment and how they work, to their role or what they have to achieve.*

In the example of the pencil that we discussed earlier, let us examine failure of the third function, that is,

- It is too thin or fat to hold, or
- It has a cross-section that is irregular or difficult to grip, or
- It is too short.

We then break down the main function into sub-functions.In the case of the pizza business, the sub-functions would be as follows:

- A purchasing system that will ensure that raw materials are fresh (for example, by arranging that meat and produce are purchased daily);
- A food preparation system suitable for making consistently high quality pizzas within 10 minutes of order;
- A communication system that will ensure voice contact with key staff, customers, and suppliers during working hours;
- A delivery system that will enable customers within a range of 10 km to receive their hot pizzas from pleasant agents within 30 minutes of placing the orders;

Each of the sub-functions can now be broken down, and we take the delivery system as an example:

- To deliver up to 60 hot (50-55°C) pizzas per hour during non-peak hours, and up to 120 hot pizzas per hour from 5:30 p.m. to 8:00 p.m.;
- To arrange deliveries such that agents do not backtrack, and that every customer is served within 30 minutes of order;
- To ensure that agents greet customers, smile, deliver the pizzas, and collect payments courteously.

These clear definitions of requirements enable the analyst to determine the success or failure of the system quite easily. The IDEF methodology promotes such clarity, and Figure 2.2 shows the Level 1 FBD of the pizza delivery system. Note that we have not thus far talked about equipment used, only what they have to do to satisfy their functional requirements. For example, the agents could be using bicycles, scooters, motorcycles, or cars to do their rounds. Similarly, they may use an insulated box to carry the pizzas, or they may use some other equipment. The only requirement is that the pizzas are delivered while they are still hot. We can break this down to show the sub-functions, as shown in Figure 2.3. Note that the inputs, outputs, controls and facilities/equipment retain their original alignment, though they may now be connected to some of the sub-function boxes.

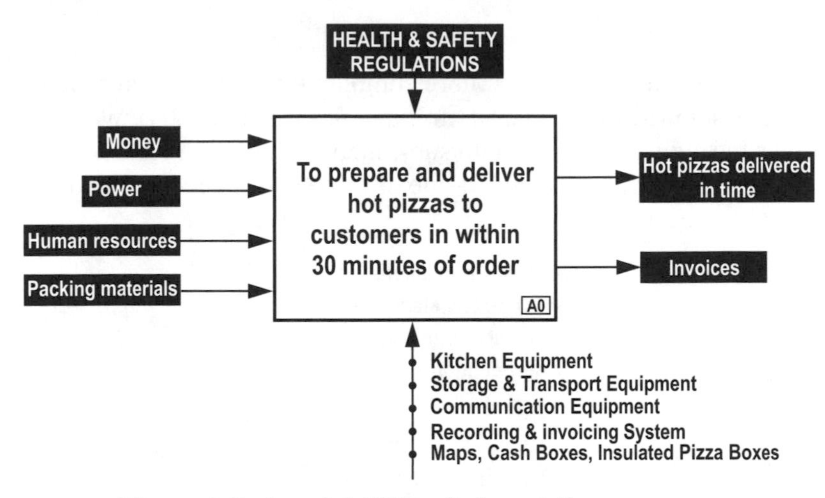

Figure 2.2 Level 0 FBD of pizza delivery system.

Figure 2.3 shows the inter-relationship between the sub-systems.

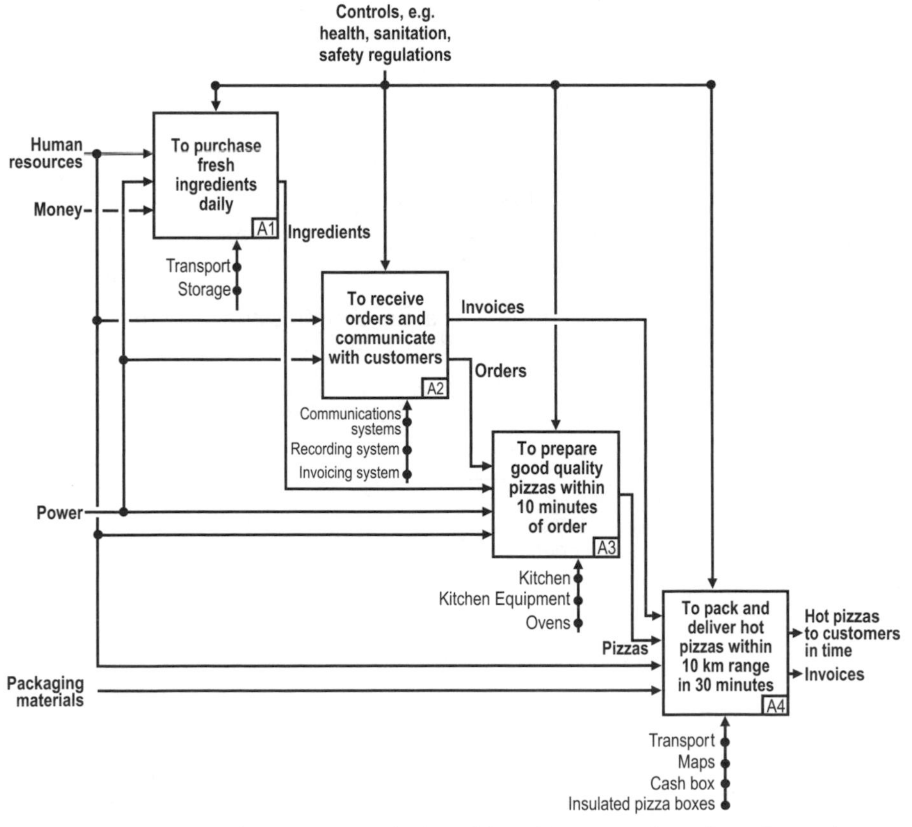

Figure 2.3 Level 1 FBD—Relationship of intermediate functions pizza delivery system.

We are now ready to address more complex industrial systems, and use a gas compressor in a process plant as an example (see Figure 2.4). We have broken down the main function A_0 into sub-functions A_1, A_2, A_3 and A_4 in figure 2.5. Thereafter, we have expanded one of these sub-functions A_2 further, as illustrated in figure 2.6.

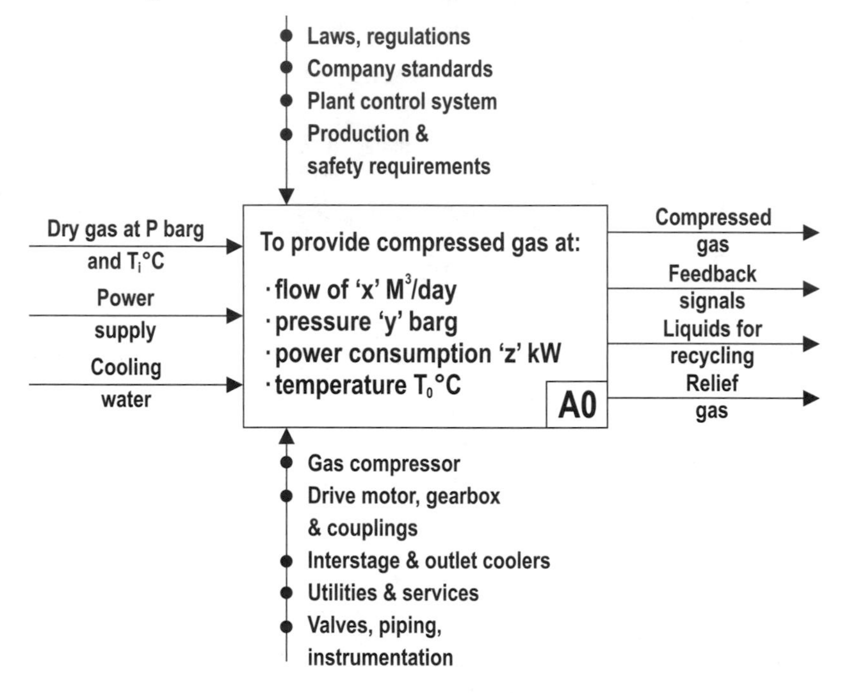

Figure 2.4 Level 0 FBD of a gas compression system.

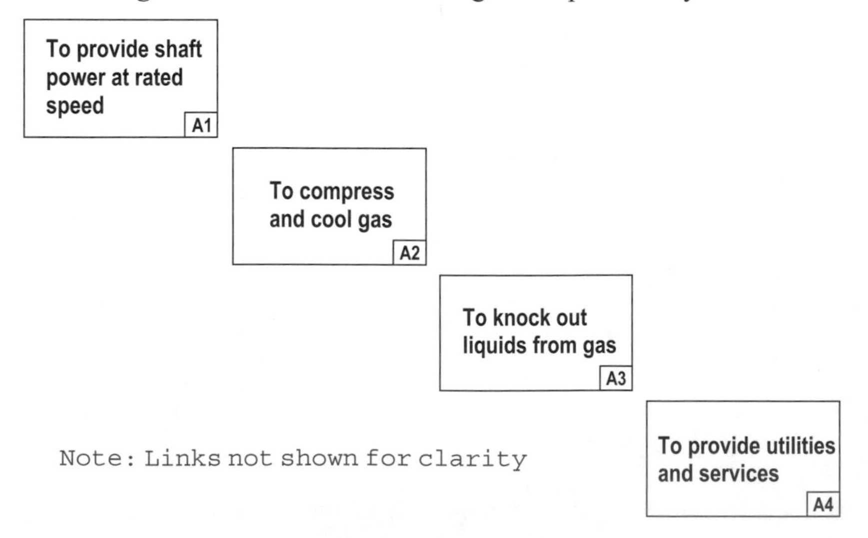

Figure 2.5 Level 1 identification of sub-system.

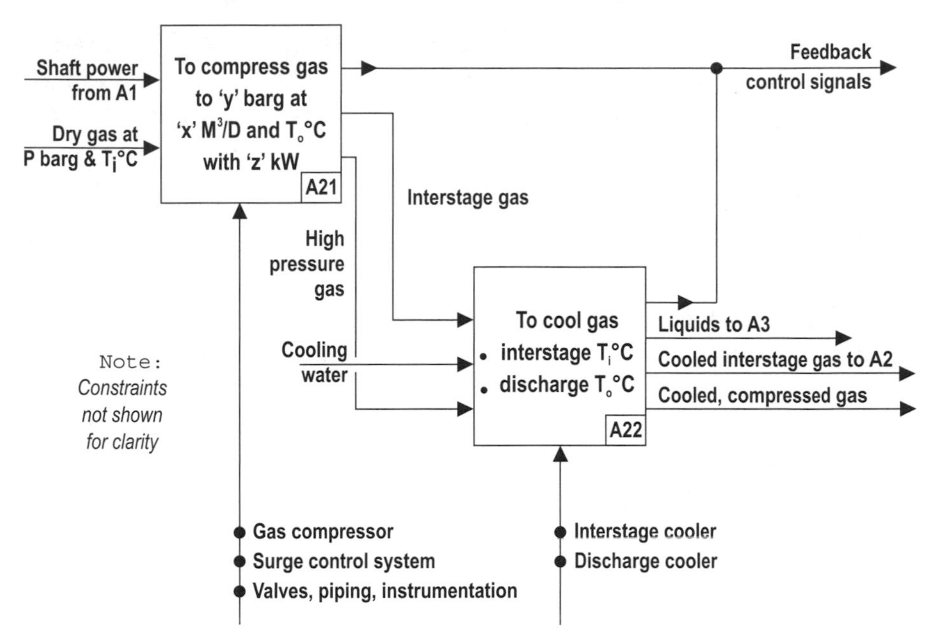

Figure 2.6 Level 2 FBD of a gas compression sub-system.

The method is applicable to any business process. We can use an FBD to describe an industrial organization, a supermarket chain, the police force, or a pizza franchise. The diagram itself may appear complex at first sight, but after some familiarization it becomes easier. The clarity and definition it brings makes it a good communication tool.

2.3 FAILURE MODES AND EFFECTS ANALYSIS (FMEA)

The performance standards embedded in the definition of the function allows identification of the success or failure of each of the systems or sub-systems. If there is a failure to achieve the objective, it is possible to identify how exactly this happens. In doing so, we identify the mode of failure. Each failure may have several failure modes.

As an example, consider engine-driven emergency generators. An important function is that they must start if the main power supply fails. They have other functions, but let us focus on this one for the moment. What are the causes of its failure to start and how can it happen? We have to establish fuel supply and combustion air, and crank the engine up in order to start it. Several things may prevent the success of the cranking operation. These include weak batteries or problems with the starter motor or the starting-clutch mechanism. If any of these failures occurs, the engine will not be able to start. These are called failure modes.

We can take this type of analysis down to a lower level. For example, the clutch itself may have failed due to a broken spring. At what level should we stop the analysis? This depends on our maintenance policy. We have the following options:

- Replace the clutch assembly, or
- Open the clutch assembly at site and replace the main element damaged, for example, the broken spring.

We can carry out the FMEA at a sub-system functional level, for example, fails to start or stopped while running, as discussed above. It is also feasible to do an FMEA at a level of the smallest replaceable element, such as that of the clutch spring. When designing process plants, a functional approach is generally used. When designing individual equipment, the manufacturers usually carry out FMEAs at the level of the non-repairable component parts. This enables the manufacturer to identify potential component reliability problems and eliminate them at the design stage. Davidson[4] gives examples of both types of FMEA applications.

In a functional analysis, we identify maintenance significant items, failures of which can cause loss of system or sub-system function. In this case, we stop the analysis at assembly level because we will replace it as a unit, and not by replacing, for example, its broken spring. Unlike the manufacturers, we cannot usually justify analysis at the lower level, because the cost of analysis will exceed the benefit. The volume of work in a component level FMEA is much higher than in a functional FMEA.

For each failure mode, there will be some identifiable local effect. For example, an alarm light may come on, or the vibration or noise level may rise. In addition there can be some effect at the overall system level. If the batteries are weak, the cranking speed will be slow, and there will be a whining noise; this is the local effect. The engine will not start, and emergency power will not be available. This may impair safety in the installation, leading to asset damage, injury or loss of life; this is the system effect.

We can identify how significant each failure mode is by examining the system effects. In this case, failure to start can eventually cause loss of life. However, if we have another power source, say a bank of batteries, the failure to start of the engine will not really matter.There may be some inconvenience, but there is no safety implication. The failure is the same; that is, the engine does not start, but the consequences are different.

The purpose of maintenance is to ensure that the system continues to function. How we maintain each sub-system will depend on the consequences, as described by the system effects. For example, if the failure of an item does not cause immediate loss of function, we can limit the maintenance to repairing it after failure. In each situation, the outcome is dependent on the configuration of the facility. The operating context may differ in seemingly identical facilities. The FBD and FMEA will help identify these differences and take the guesswork out of decision making.

2.4 EFFECTIVE PLANNING

The elegance of the functional approach will now be clear. For every business, we can define its objectives at the top level, or its overall functions. We can break these down to identify the related systems and sub-systems. Next, we identify the functions of each system and sub-system, and carry out an FMEA. The analysis is applicable to an operating plant or to one that is still on the drawing board. As a result of this top-down approach, we can concentrate the planning effort on what really matters to the organization.

Individuals and organizations can fall into the trap of rewarding activity instead of the results achieved. Movement and activity are often associated with hard work. Sometimes this is of no avail, so activity by itself has no merit. We have to plan the work properly so as to achieve meaningful results.

The functional analysis concentrates on the results obtained, and the quality standards required. We have discussed its use in the context of maintenance work, but we can apply the method in any situation where we can specify the results clearly. For example, Knotts[1] discusses their use in the context of business process re-engineering.

2.5 PREVENTION OF FAILURES OR MITIGATION OF CONSEQUENCES?

Once we identify the functional failures, the question arises as to how best to minimize their impact. Two solutions are possible: 1) we can try to eliminate or minimize the frequency of failures or 2) take action to mitigate the consequences.

If we can determine the root cause of the failure, we may be able to address the issue of frequency of events. Usually, this will mean elimination of the root cause. Historically, human failures have accounted for nearly three quarters of the total. Hence, merely designing stronger widgets will not always do the trick. Not doing the correct maintenance on time to the right quality standards can be the root cause, and this is best rectified by re-training or addressing a drop in employee motivation Similarly, changes in work practices and procedures may eliminate the root cause. All of these steps, including physical design changes are all considered a form of redesign. In using these methods, we are attempting to improve the intrinsic or operational reliability of the equipment, sub-system, or system. As a result, we expect to see a reduction in the failure rate or frequency of occurrence.

An alternative approach is to accept the failure rates as they are, and devise a method to reduce their consequences. The aim is to do the applicable and effective maintenance tasks at the right time, so that the consequences are minimal. We will discuss both of these risk reduction methods, and the tools we can use, in Chapter 10.

Once we identify the tasks, we schedule the tasks, arranging the required resources, materials, and support services. Thereafter, we execute the work to the correct quality standards. Lastly, we record and analyze the performance data, to learn how to plan and execute the work more effectively and efficiently in the future.

When there are safety consequences, the first effort must be to reduce the exposure, by limiting the number of people at risk. Only those people who need to be there to carry out the work safely and to the right quality standards should be present. Maintaining protective devices so that they operate when required, is also important. Should a major incident take place in spite of all efforts, we must have damage limitation procedures, equipment designed to cope with such incidents, and people trained in emergency response. A recent example showed the usefulness of such damage limitation preparedness. In September 1997, an express train traveling from Swansea to London crashed into a freight train, at Southall, just a few miles before reaching London-Paddington station. The freight train was crossing the path of the passenger train, which was traveling at full speed, so one can visualize the seriousness of the accident. The response of the rescue and emergency services was excellent. The prompt and efficient rescue services should be given full credit as the death toll could have been considerably worse than the seven fatalities that occurred.

2.6 CHAPTER SUMMARY

The functional approach is aligned closely with the objectives of a business. The IDEF methodology is an effective way to understand and communicate this approach. We used this tool to understand the functions of a range of applications, from pencils and pizza business to gas compression systems in process plants. A clear definition of the functions enables us to identify and understand functional failure. Thereafter, we use the FMEA to analyze functional failures. We make a distinction between the use of the functional and equipment level FMEAs. Using a top-down approach, we identify functional failures and establish their importance.

In managing risks, we can try to reduce the frequency of failure or mitigate their consequences. Both methods are applicable, and the applicability, effectiveness, and the cost of doing one or the other will determine the selection. Lastly, we touched on the importance of damage limitation measures.

References

1) Knotts, Rob. 1997. "Asset Maintenance Business Management: A need and An Approach. Exeter: Proceedings, 7th International M.I.R.C.E. Symposium on System Operational Effectiveness. 129-133.
2) Mayer, R.J. 1994. "IDEFO Functional Modeling: A Reconstruction of the Original." Air Force Wright Aeronautical Laboratory Technical Report. Knowledge Based Systems Inc. AFWAL-TR-81-4023.
3) Jones, R.B. 1995. *Risk-Based Management:A Reliability-Centered Approach.* Gulf Professional Publishing Company. ISBN: 0884157857.
4) Davidson J. 1994. *The Reliability of Mechanical Systems*. Mechanical Engineering Publications Ltd. ISBN: 0852988818. 78-82.

Further reading

1) Anderson R.T., and L.Neri. 1990. *Reliability Centered Maintenance: Management and Engineering*. Elsevier Applied Science Publishers, Ltd. ASIN: 185166470X.
2) Smith A.M. 1993. *Reliability Centered Maintenance*. Butterworth-Heinemann. AISN: 007059046X
3) Moubray J. 2001. *Reliability-Centred Maintenance.* Industrial Press, Inc. ISBN: 0831131462.

CHAPTER 3

Reliability Engineering for the Maintenance Practitioner

We can now develop some of the reliability engineering concepts that we will need in subsequent chapters. Prior knowledge of the subject is not essential, as we will define the relevant terms and derive the necessary mathematical expressions. As this is not a text on reliability engineering, we will limit the scope of our discussion to the following areas of interest.

- Failure histograms and probability density curves;
- Survival probability and hazard rate;
- Constant hazard rates, calculation of test intervals, and errors with the use of approximations;
- Failure distributions and patterns, and the use of the Weibull distribution;
- Generation of Weibull plots from maintenance records;
- Weibull shape factor and its use in identifying maintenance strategies;

For a more detailed study of reliability engineering, we suggest that readers refer to the texts[3,4,6] listed at the end of the chapter.

3.1 FAILURE HISTOGRAMS

We discussed failures at the system level in Chapter 2. These are as a result of one or more modes of failure at the component level. In the example of the engine's failure to crank up, we identified three of the failure modes that may cause the failure of the cranking mechanism.

If designers and manufacturers are able to predict the occurrence of these failures, they can advise the customers when to take corrective actions. With this knowledge, the customers can avoid unexpected production losses or safety incidents. Designers also require this information to improve the reliability of their products. In the case of mass-produced items, the manufacturer can test representative samples from the production line and estimate their reliability performance. In order to obtain the results quickly, we use accelerated tests. In these tests, we subject the item to higher stress levels or operate it at higher speeds than normal so that this initiates failure earlier.

Let us take as an example the testing of a switch used in industrial applications. Using statistical sampling methods, the inspector selects a set of 37 switches from a given batch, to assess the life of the contacts. These contacts can burn out, resulting in the switch failing to close the circuit when in the closed position. In assessing the time-to-failure of switches, a good measure is the number of operations in service. The test consists of repeatedly moving the switch between the on and off positions under full load current conditions. During the test, we operate the switch at a much higher frequency than expected normally.

As the test progresses, the inspector records the failures against the number of operations.When measuring life performance, time-to-failure may be in terms of the number of cycles, number of starts, distance traveled, or calendar time. We choose the parameter most representative of the life of the item. In our example, we measure 'time' in units of cycles of tests. The test continues till all the items have failed. In Table 3.1, a record of the switch failures after every thousand cycles of operation is shown.

Cycles	Failures	Cumulative	% Cumulative	Survivors	No. Failed/Sample Size
1000	0	0	0.00%	37	0
2000	1	1	2.70%	36	0.027027
3000	3	4	10.81%	33	0.081081
4000	6	10	27.03%	27	0.162162
5000	7	17	45.95%	30	0.189189
6000	7	24	64.86%	13	0.189189
7000	6	30	81.08%	7	0.162162
8000	4	34	91.89%	3	0.108108
9000	2	36	97.30%	1	0.154054
10000	1	37	100.0%	0	0.027027
Total	**37**				

Table 3.1

We can plot this as a bar chart (see Figure 3.1), with the number of switch failures along the Y axis, and the life measured in cycles along the X axis.

To find out how many switch failures occurred in the first three thousand cycles, we add the corresponding failures, namely $0+1+3=4$. By deducting the cumulative failures from the sample size, we obtain the number of survivors at this point as 37 - 4 = 33. As a percentage of the total number of

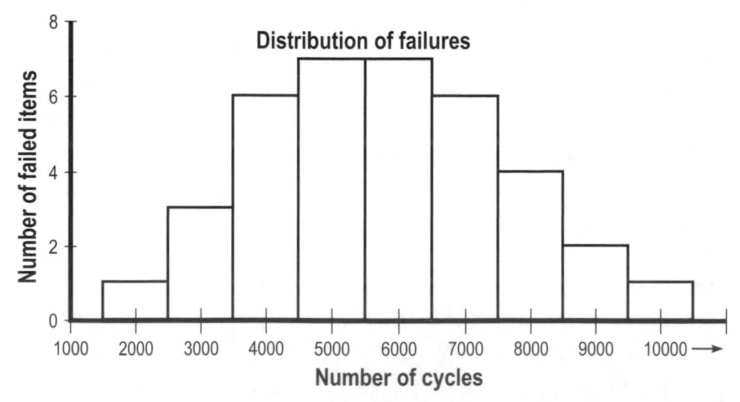

Figure 3.1 Number of failures recorded per cycle.

recorded failures, the corresponding figures are 4/37 or approximately 11% and 33/37 or approximately 89% respectively.

We can view this information from a different angle. At the end of three thousand cycles, about 11% of the switches have failed and 89% have survived. Can we use this information to predict the performance of a single switch? We could state that a switch that had not failed during the first three thousand cycles, had a survival probability of approximately 89%. Another way of stating this is to say that the reliability of the switch at this point is 89%. There is no guarantee that the switch will last any longer, but there is an 89% chance that it will survive beyond this point. As time passes, this reliability figure will keep falling. Referring to the table 3.1, we can see that at the end of five thousand cycles,

- The cumulative number of failures is 17;
- The proportion of cumulative failure to the sample size (37) is 46%;
- The proportion of survivors is about 100% - 46% = 54%.

In other words, the reliability is about 54% at the end of five thousand cycles. Using the same method, by the end of nine thousand cycles the reliability is less than 3%.

How large should the sample be, and will the results be different with a larger sample? With a homogeneous sample, the actual percentages will not change significantly, but the confidence in the results increases as the sample becomes larger. The cost of testing increases with the sample size, so we have to find a balance and get meaningful results at an acceptable cost. With a larger sample, we can get a better resolution of the curve, as the steps will be smaller and the histogram will approach a smooth curve. We can normalize the curve by dividing the number of failures at any point by the sample size, so that the height of the curve shows the failures as a ratio of the sample size. The last column of Table 3.1 shows these normalized figures.

3.2 PROBABILITY DENSITY FUNCTION

This brings us to the concept of probability density functions. In the earlier example, we can smooth the histogram in Figure 3.1 and obtain a result as seen in Figure 3.2. The area under the curve represents the 37 failures, and is normalized by dividing the number of failures at any point by 37, the sample size. In reliability engineering terminology, we call this normalized curve a probability density function or ***pdf*** curve. Since we tested all the items in the sample to destruction, the ratio of the total number of failures to the sample size is 1. The total area under the ***pdf*** curve represents the proportion of cumulative failures, which is also 1.

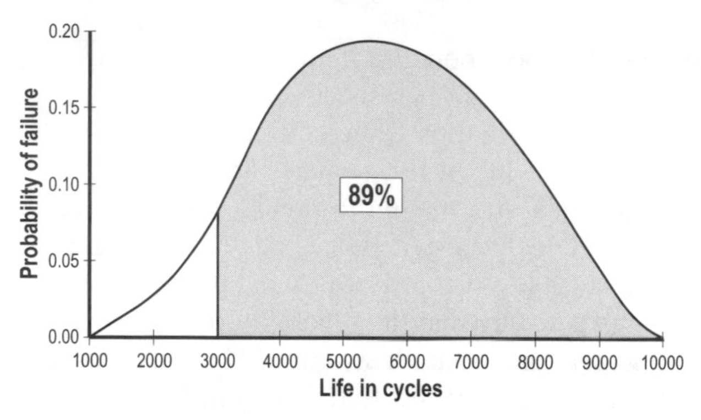

Figure 3.2 Probability density function.

If we draw a vertical line at time t = 3,000 cycles, the height of the curve gives the number of failures as a proportion to the sample size, at this point in time. The area to the left of this line represents the cumulative failure probability of 11%, or the chance that 4 of the 37 items would have failed. The area to the right represents the survival probability of 89%. In reliability engineering terminology, the survival probability is the same as its reliability, and the terms are interchangeable.

3.3 MORTALITY

We now turn to the concept of mortality, which when applied in the human context, is the ratio of the number of deaths to the surviving population. To illustrate this, let us consider the population in a geographical area. Let us say that there are 100,000 people in the area on the day in question. If there were ten deaths in all on that day, the mortality rate was 10/100,000, or 0.0001.
Actuaries analyze the mortality of a population with respect to their age. They measure the proportion of the population that die within one, two, three ....n years. A plot of these mortality values would be similar to Figure 3.3, Element A (which refers to equipment component failures). In the first part of the curve (the so-called infant mortality section) the mortality rate keeps falling. A baby has a high chance of dying at birth, and the longer it survives, the greater the chance that it will continue to live. After the first few days or weeks, the mortality rate levels out, and for the next 50-70 years, it is fairly constant. People die randomly, due to events such as road accidents, food poisoning, homicides, cancer, heart disease, or other reasons. Depending on their lifestyles, diet, race, and sex, from about 50 years on the mortality starts to rise. As they get older, they become susceptible to more diseases, their bones tend to become brittle, and their general resistance becomes lower. Not many people live up to 100 years, though some ethnic groups have

exceptional longevity. Insurance companies use these curves to calculate their risks. They adjust the premiums to reflect their assessment of the risks.

We use a similar concept in reliability engineering. The height of the ***pdf*** curve gives the number of failures at any point in time, and the area of the curve to the right of this point the number of survivors. The term ***hazard rate*** designates ***equipment mortality***. We divide the number of failures by the number of survivors, at this point. In the earlier example, the hazard rate at t = 3,000 cycles is 4/33 or 0.12. The study of hazard rates gives us an insight into the behavior of equipment failures, and enables us to make predictions about future performance.

3.4 HAZARD RATES AND FAILURE PATTERNS

The design of industrial equipment was simple, sturdy, heavy, and robust prior to World War II. Repairs were fairly simple, and could easily be done at site using ordinary hand tools. Breakdown strategies were common, which meant that equipment operated till failures occurred. The introduction of mass production techniques meant that interruptions of production machinery or conveyors resulted in large losses. At the same time, the design of equipment became more complex. Greater knowledge of materials of construction led to better designs with a reduction in weight and cost. Computer-aided analysis and design tools became available, along with computing capacity. As a result, the designers could reduce safety factors (which included a factor for uncertainty or ignorance). In order to reduce investment costs, designers reduced the number of standby equipment installed and intermediate storage or buffer stocks.

These changes resulted in slender, light, and sleek machinery. They were not as rugged as their predecessors, but met the design conditions. In order to reduce unit costs, machine uptime was important. The preferred strategy was to replace complete sub-assemblies as it took more time to replace failed component parts.

A stoppage of high volume production lines resulted in large losses of revenue. In order to prevent such breakdowns, manufacturers used a new strategy. They replaced the sub-assemblies or parts at a convenient time before the failures occurred, so that the equipment was in good shape when needed. The dawn of planned preventive maintenance had arrived.

Prior to the 1960s, people believed that most failures followed the so-called bath-tub curve. This model is very attractive, as it is so similar to the human mortality curves. By identifying the knee of the curve, namely, the point where the flat curve starts to rise, one could determine the timing of maintenance actions. Later research[1] showed that only a small proportion of component failures followed the bath-tub model, and that the constant hazard pattern accounted for the majority of failures. In the cases where the bath-tub model did apply, finding the knee of the curve is not a trivial task.

As a result, conservative judgment prevailed when estimating the remaining life of components. Preventive maintenance strategies require that we replace parts before failure, so the useful life became synonymous with the shortest recorded life. Thus the replacement of many components took place long before the end of their useful life. The opportunity cost of lost production justified the cost of replacing components that were still in good condition.

The popularity of preventive maintenance grew especially in industries where the cost of downtime was high. This strategy was clearly costly, but was well justified in some cases. However, the loss of production due to planned maintenance itself was a new source of concern. Managers who had to reduce unit costs in order to remain profitable started to take notice of the production losses and the rising cost of maintenance.

Use of steam and electrical power increased rapidly throughout the twentieth century. Unfortunately there were a large number of industrial accidents associated with the use of steam and electricity. This resulted in the introduction of safety legislation to regulate the industries. At this time, the belief was that all failures were age related, so it was appropriate to legislate time-based inspections. They felt that the number of incidents would reduce by increasing the inspection frequencies.

Intuitively, people felt more comfortable with these higher frequency inspection regimes. Industrial complexity increased greatly from the mid-1950s onwards with the expansion of the airline, nuclear, and chemical industries. The number of accidents involving multiple fatalities experienced by these industries rose steeply.

By the late 1950s, commercial aviation became quite popular. The large increase in the number of commercial flights resulted in a corresponding increase in accidents in the airline industry. Engine failures accounted for a majority of the accidents and the situation did not improve by increasing maintenance effort. The regulatory body, the U.S. Federal Aviation Agency decided to take urgent action in 1960, and formed a joint project with the airline industry to find the underlying causes and propose effective solutions.

Stanley Nowlan and Howard Heap[1], both of United Airlines, headed a research project team that categorized airline industry failures into one of six patterns. The patterns under consideration are plots of hazard rates against time. Their study revealed two important characteristics of failures in the Airline Industry, hitherto unknown or not fully appreciated.

1. The failures fell into six categories, illustrated in Figure 3.3.
2. The distribution of failures in each pattern revealed that **only 11% were age-related**. The **remaining 89%** appeared to have failures **not related to component age.** This is illustrated in the pie-chart, Figure 3.4.

The commonly held belief that all failures followed Pattern A-the Bathtub Curve, justifying age-based preventive maintence was called into question, as

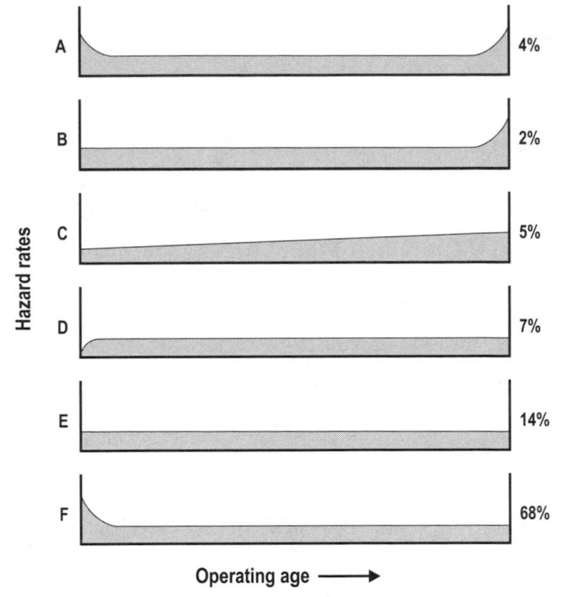

Figure 3.3 Failure Patterns

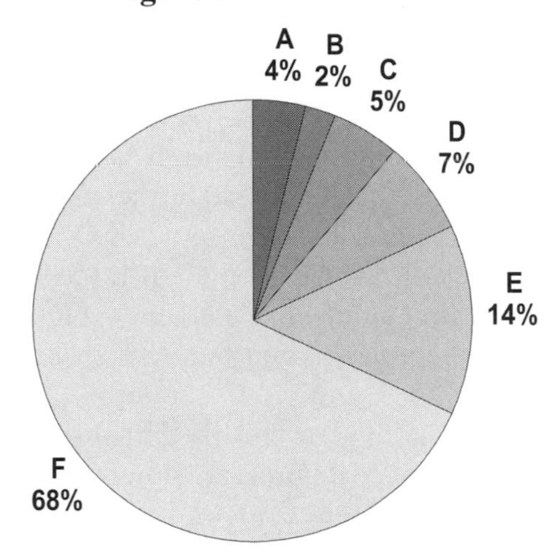

Figure 3.4 Faliure Patterns

Patterns A, B, and C, which are age-related, account for 11% of failures studied in research project

it accounted for just a small percentage of all failures (in the Airline Industry). Nowlan and Heap questioned the justification for doing all maintenace using age as the only criterion.

We will discuss these issues later in the book.

An explanation of these failure patterns and a method to derive them using a set of artificially created failure data is given in Appendix 3-1.

3.5 THE TROUBLE WITH AVERAGES

As we know, the average height of a given population does not tell us a great deal. If the average is, say, 1.7 m, we know that there will be some people who are shorter, say under 1.5 m, and some who are taller, perhaps over 2 m. If you are a manufacturer of clothes, you would need to know the spread or distribution of the heights of the population in order to design a range of sizes that are suitable.

We use the average or mean as a measure to describe a set of values. The arithmetic average is the one most commonly used, since it is easy to compute. The term average may give the impression it is an expected value. In practice, these two values may be quite different from each other.

There is a similar situation when we deal with equipment failure rates. The majority of the failures may take place in the last few weeks of operation, thereby skewing the distribution. For example, if we recorded failures of 100 tires, and their combined operational life was three million km, what can we learn from the mean value of 30,000 km of average operational life? In practice, it is likely that there were very few failures within the first 5000 km or so, and that a significant number of tires failed after 30,000 km. Hence the actual distribution of failures is important if we are to use this information for predicting future performance. Such predictions are useful in planning resources, ordering replacement spares, and in preparing budgets.

As a refinement, we can define the spread further using the standard deviation. However, even this is inadequate to describe the distribution pattern itself, as illustrated by the following example. In Table 3.2, you can see three sets of failure records of a set of machine elements. Figures 3.5, 3.6, and 3.7 respectively illustrate the corresponding failure distributions, labeled P, Q, and R.

Note that all three distributions have nearly the same mean values and standard deviations. The failure distributions are however quite different. Most of the failures in distribution P are after about 5 months, while in distribution R, there are relatively few failures after 20 months. Thus the two distributions are skewed, one to the left and the other to the right. The distribution Q is fairly symmetrical. *Knezevic*[2] *discusses the importance of knowing the actual distribution in some detail. He concludes his paper with the following observations.*

- The knowledge of the actual failure distribution can be important;
- The use of a constant failure rate is not always appropriate;
- As investment and operational expenditure gets greater scrutiny, the pressure to predict performance will increase—in many cases, the use of mean values alone will reduce the accuracy of predictions;
- Understanding the distributions does not need more testing or data.

	Distribution of failures		
	Element		
Month	P	Q	R
1	1	0	0
2	1	0	1
3	1	1	10
4	2	2	22
5	2	5	31
6	3	9	34
7	4	15	36
8	6	23	36
9	8	28	35
10	11	32	33
11	15	33	30
12	19	35	24
13	24	36	19
14	30	34	15
15	33	31	11
16	35	25	8
17	36	23	6
18	36	16	4
19	34	9	3
20	31	5	2
21	22	2	2
22	10	1	1
23	1	0	1
24	0	0	1
Total	365	365	365
Mean	15.20833	15.20833	05.20833
Std.Dev.	13.79344	13.78398	13.79344

Table 3.2 Distribution of failures - elements P, Q & R.

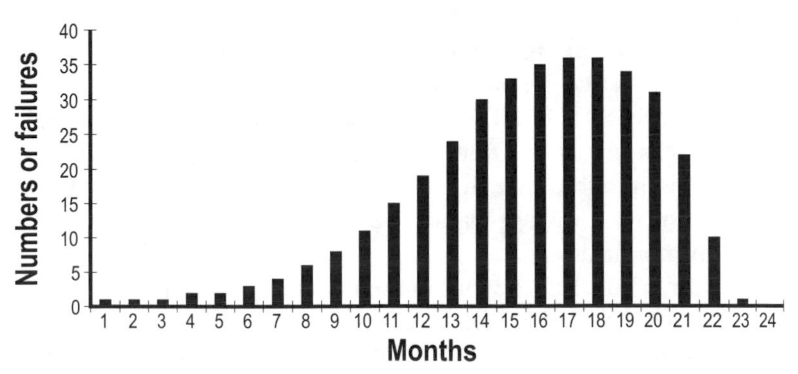

Figure 3.5 Distribution P : Mean = 15.21. Std.Dev.= 13.79

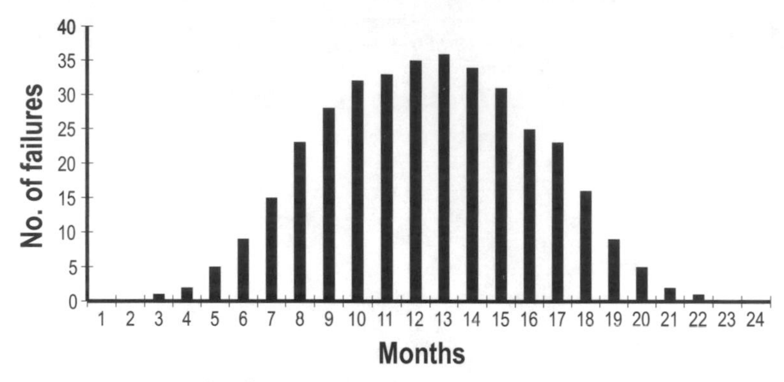

Figure 3.6 Distribution Q : Mean = 15.21. Std.Dev.= 13.79

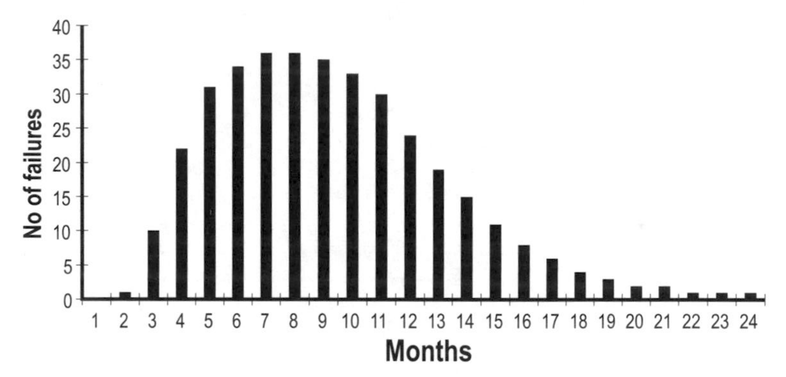

Figure 3.7 Distribution R : Mean = 15.21. Std.Dev.= 13.79

3.6 THE SPECIAL CASE OF THE CONSTANT HAZARD RATE

So far we have emphasized the importance of knowing the actual failure distribution. One should not assume a constant or average failure rate, unless there is evidence to believe this to be true. However, we know that in the airline industry, of the six patterns (Figures 3.3), the patterns D, E and F account for about 89% of the failures. Patterns D and F are similar to pattern E over most of the life. If we ignore early failures, the constant hazard pattern accounts for nearly 89% of all the failures. The picture is similar in the offshore oil and gas industry. Other industries have not published their own results, but it is possible that they have somewhat similar distributions as well. In view of its dominant status, the special case of the constant hazard rate merits further discussion.

Let us examine the underlying mathematical derivations relating to constant hazard rates. In section 3.3, we defined the hazard rate as the ratio of the probability of failure at any given time to the probability of survival at that time. We can express this using the following equation.

$$z(t) = \frac{f(t)}{R(t)} \qquad 3.1$$

where ***z(t)*** is the hazard rate, ***f(t)*** is the probability of failure, or the height of the ***pdf*** curve, and ***R(t)*** is the survival probability, or the area of the ***pdf*** curve to the right, at time ***t***. The cumulative failure is the area of the curve to the left at time ***t***. The total area under the ***pdf*** curve, that is, cumulative failures plus survivors has to be 100% or 1.

$$F(t) + R(t) = 1 \qquad 3.2$$

and

$$F(t) = \int_0^t f(t)dt \qquad 3.3$$

or

$$f(t) = \frac{dF(t)}{dt}$$

hence

$$f(t) = \frac{d\{1 - R(t)\}}{dt} = \frac{-dR(t)}{dt} \qquad 3.4$$

The constant hazard rate will be denoted as λ, and is given by

$$\frac{f(t)}{R(t)} = \lambda \qquad 3.5$$

Combining expressions *3.4* and *3.5*, we get,

$$\lambda \times R(t) = \frac{-dR(t)}{dt}$$

or

$$-\lambda = \frac{1}{R(t)} \times \frac{dR(t)}{dt}$$

Integrating,

$$e^{-\lambda t} = R(t) \text{ for } t > 0 \qquad 3.6$$

3.7 AVAILABILITY

Availability is a measure of the time an equipment is able to perform to specified standards, in relation to the time it is in service. The item will be unable to perform when it is down for planned or unplanned maintenance, or when it has tripped. Note that it is only required that the equipment is able to operate, and not that it is actually running. If the operator chooses not to operate it, this does not reduce its availability.

Some items are only required to operate when another item fails, or a specific event takes place. If the first item itself is in a failed state, the operator will not be aware of its condition since it is not required to work till another event takes place. *Such failures are called hidden failures. Items subject to hidden failures can be in a failed state any time after installation, but we will not be aware of this situation.* The only way to know if the item is working is to place a demand on it. For example, if we want to know whether a firc pump will start, it must be actually started - this can be by a test or if there is a real fire. At any point in its life, we will not know whether it is in working condition or has failed. If it has failed, it will not start. The survival probability gives us the expected value of its up-state, and hence its availability on demand at this time. *Thus, the availability on demand is the same as the probability of survival at any point in time.* This will vary with time, as the survival probability will keep decreasing, and with it the availability. This brings us to the concept of mean availability.

3.8 MEAN AVAILABILITY

If we know the shape of the ***pdf*** curve, we can estimate the item's survival probability. If the item has not failed till time **t**, the reliability function **R(t)** gives us the probability of survival up to that point. As discussed above, this is the same as the instantaneous availability.

In the case of hidden failures, we will never know the exact time of failure. We need to collect data on failures by testing the item under consideration periodically. It is unlikely that a single item will fail often enough in a test situation to be able to evaluate its failure distribution. So we collect data from several similar items operating in a similar way and failing in a similar manner, to obtain a larger set (strictly speaking, all the failures must be independent and identical, so using similar failures is an approximation). We make a further assumption, that the hazard rate is constant. When the hazard rate is constant, we call it the *failure rate.* The inverse of the failure rate is the Mean Time To Failures or MTTF. MTTF is a measure of average operating performance for non-repairable items, obtained by dividing the cumulative time in service (hours, cycles, miles or other equivalent units) by the cumulative number of failures. By non-repairable, we mean items that are replaced as a whole, such as light bulbs, ball bearings or printed circuit boards. In the case or repairable items, a similar measure of average operating performance is used, called Mean Operating Time Between Failures, or MTBF. This is obtained by dividing the cumulative time in service (hours, cycles, miles or other equivalent units) by the cumulative number of failures. If after each repair, the item is as good as new (AGAN), it has the same value as MTTF. In practice the item may not be AGAN in every case. In the rest of this chapter, we will use the term MTBF to represent both terms.

Another term used in a related context is Mean Time to Restore, or MTTR. This is a measure of average maintenance performance, obtained by dividing the cumulative time for a number of consecutive repairs on a given repairable item (hours) by the cumulative number of failures of the item. The term 'restore' means the time from when the equipment was stopped to the time the equipment was restarted and operated satisfactorily.

Table 3.3 shows a set of data describing failure pattern E. here we show the surviving population at the beginning of each week instead of that at the end of each week. Figure 3.8 shows the cumulative number of failures, and Figure 3.9 shows the surviving population at the beginning of the first 14 weeks.

Constant Hazard Rate Data			
At start of Week No.	**Failures in prior week**	**Cumulative failures**	**Survivors from sample**
1	0	0	1000
2	15	15	985
3	15	30	970
4	14	44	956
5	14	58	942
6	14	72	928
7	14	86	914
8	14	100	900
9	13	113	887
10	13	126	874
11	13	139	861
12	13	152	848
13	12	165	835
14	12	177	823
Note: Initial sample size is 1000. Column 2 shows failures in previous week. Cumulative failures are obtained by adding failures from Week 1 to date. Survivors are obtained by deducting cumulative failures from initial sample size.			

Table 3.3

We can use this constant slope geometry in Figure 3.8 to calculate the MTBF and failure rates. When there are many items in a sample, each with a different service life, we obtain the MTBF by dividing the cumulative time in operation by the total number of failures. We obtain the failure rate by dividing the number of failures by the time in operation. Thus,

$$MTBF = \frac{1}{\lambda} \qquad 3.7$$

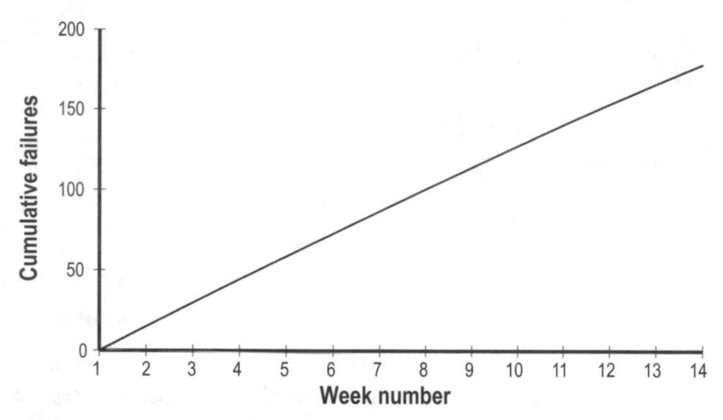

Figure 3.8 Cumulative failures against elapsed time

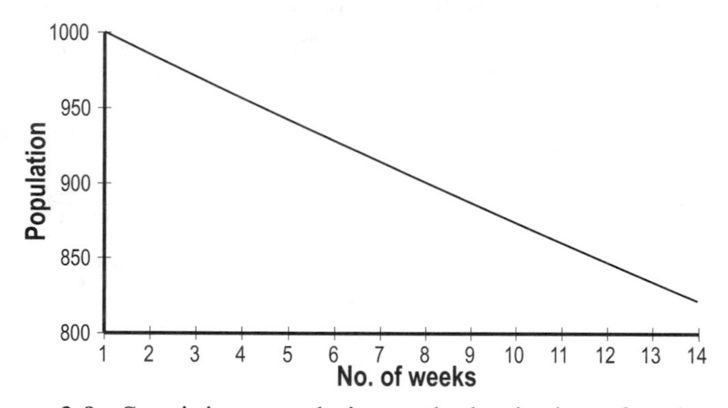

Figure 3.9 Surviving population at the beginning of each week

For a rigorous derivation, refer to Hoyland and Rausand[3], page 31. Note that this is the only case when the relationship applies, as in the other failure distributions, the slope of the cumulative failure curve changes all the time.

We can only replace an item after a test as it is a hidden failure. We do not know if it is in a failed condition unless we try to use it. How do we determine a justifiable test interval T? At the time of test, if we find the majority of items in a failed state, we have probably waited too long. In other words, we expect very high survival probability. Thus, in the case of systems affecting safety or environmental performance, it would be reasonable to expect this to be 97.5% or more, based on, for example, a Quantitative Risk Assessment.

Let us try to work out the test interval with a numerical example. Using the data in Table 3.3 at the beginning of week number 1, all 1000 items will be in sound working order (As Good As New, or AGAN). At the beginning of week number 2, we can expect 985 items to be in working order, and 970 items at the beginning week 3. At the beginning of week 14, we can expect only 823 items to be in working condition. So far, we have not replaced any of the defective items because we have not tested them and do not know how many are in a failed state.Had we carried out a test at the beginning of week 2, we would have expected to find only 985 in working order. This is therefore the availability at the beginning of week 2. If we delay the test to the beginning of week 14, only 823 items are likely to be in working order. The availability at that time is thus 823 out of the 1000 items, or 0.823.

The mean availability over any time period, say a week, can be calculated by averaging the survival probabilities at the beginning and end of the week in question. For the whole period, we can add up the point availability at the beginning of each week, and divide it by the number of weeks. This is the same as measuring the area under the curve and dividing it by the base to get the mean height. In our example, this gives a value of 91.08%. If the test interval is sufficiently small, we can treat the curve as a straight line. Using this approximation, the value is 90.81%. The error increases as we increase the test interval, because the assumption of a linear relationship becomes less applicable. We will see later that the error using this approximation becomes unacceptable, once T/MTBF exceeds 0.2.

Within the limits of applicability, the error introduced by averaging the survival probabilities at the beginning and end of the test period is fairly small (- 0.3 %). These requirements and limits are as follows.

- They are hidden failures and follow an exponential distribution;
- The MTBF > the test interval, say by a factor of 5 or more;
- The item is as good as new at the start of the interval;
- The time to carry out the test is negligible;
- The test interval > 0.

In the example, the test interval (14 weeks) is relatively small compared to the MTBF (which is 1/0.015 or 66.7 weeks). Figure 3.10 illustrates these conditions, and the terms used.

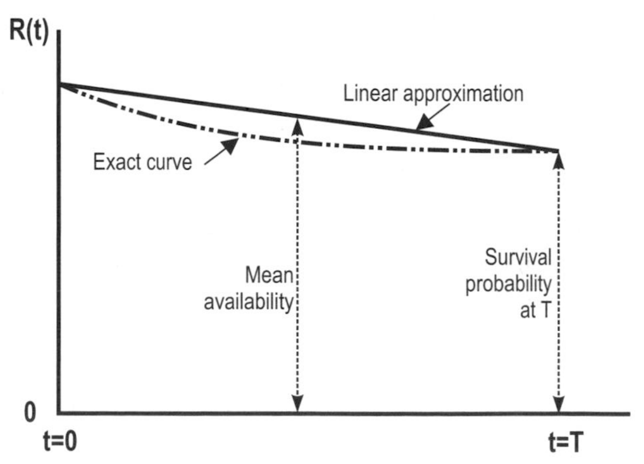

Figure 3.10 Mean availability approximation.

The objective is to have an acceptable survival probability at the time of the test. The difference in the number of survivors, calculated using the exact and approximate solutions is quite small, as can be seen in Figure 3.11. The mean availability and survival probability are related, and this is illustrated in Figure 3.12. The relationship is linear over the range under consideration.

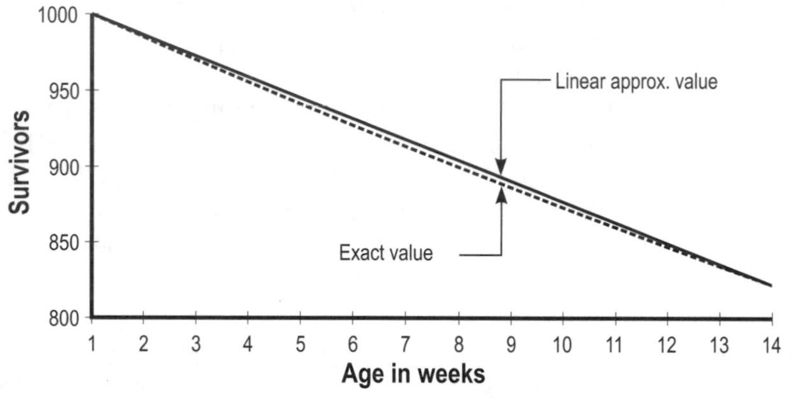

Figure 3.11 Survivors; lower curve = exact value, upper curve = linear approximation

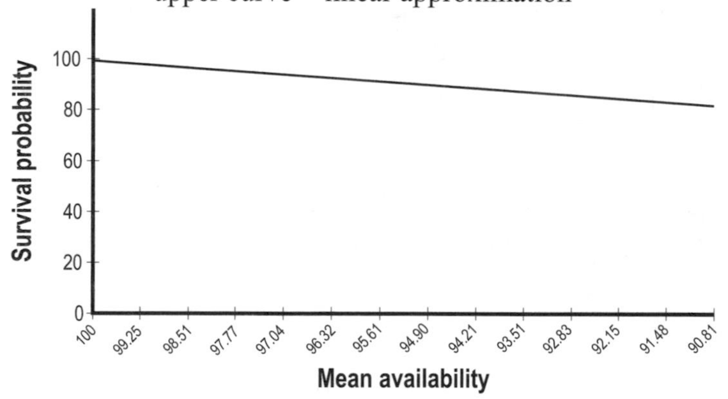

Figure 3.12 Mean availability and survival probability.

We will use this example to develop a generally applicable method to determine test intervals for hidden functions. The objective of the exercise is to find a test interval **T** that will give us the required mean availability A, when the failure rate is λ. We have noted that at any point in time, the availability of the item is the same as its survival probability, or the height of the R(t) curve. The mean availability is obtained by dividing the area of the R(t) curve by the base, thus,

$$A = \frac{1}{T} \times \int_0^T R(t)dt \qquad 3.8$$

When the hazard rate is constant, from the earlier derivation (expression 3.6),

$$for\ t > 0$$

Substituting,

$$A = \frac{1}{T} \times \int_0^T e^{-\lambda t} dt \qquad 3.9$$

Evaluating the integral explicitly gives

$$A = \frac{1}{\lambda T} \times \left(1 - e^{-\lambda T}\right) \qquad 3.10$$

This gives an exact measure of mean availability. We cannot use algebraic methods to solve the equation, as **T** appears in the exponent as well as in the main expression. We can of course use numerical solutions, but these are not always convenient, so we suggest a simpler approximation, as follows.

The survival probability or **R(t)** curve (see fig.3.10) is nearly linear over the test interval **T**, under the right conditions. The mean is the arithmetic average of the height of the curve at **t = 0** and **t = T**.

The mean value of availability A is then:

$$A \approx \frac{1}{2} \times \left(e^{-\lambda t @ t=0} + e^{-\lambda t @ t=T}\right) \qquad 3.11$$

$$A \approx 0.5 \times \left(1 + e^{-\lambda T}\right) \qquad 3.12$$

or,

$$-\lambda T \approx \ln(2A - 1) \qquad 3.13$$

The estimates produced by this expression are slightly optimistic. However over the range of applicability, the magnitude of deviation is quite small. Table 3.4 and Figure 3.11 show the error in using the exact and approximate equations for values of **λt** from 0.01 to 0.25. Figure 3.12 shows how close the approximate value is to the exact value of mean availability over the range. It is quite small up to a value of **λt** of 0.2. Therefore, we can safely use the approximation within these limits.

T/MTBF	**MTBF/T**	**exp(-T/MTBF)**	**Exact Av.**	**Approx. Av**	**Difference**
0.01	100.00	0.990049834	0.995016625	0.99502492	8.218E-06
0.02	50.00	0.960198673	0.990066335	0.99009934	3.3002E-05
0.03	33.33	0.970445534	0.985148882	0.98522277	7.3885E-05
0.05	20.00	0.951229425	0.97541151	0.97561471	2.03E-04
0.10	10.00	0.904837418	0.95162582	0.95241871	7.93E-04
0.12	8.33	0.886920437	0.942329694	0.94346022	1.13E-03
0.14	7.14	0.869358235	0.933155461	0.93467912	1.52E-03
(continued)					

T/MTBF	MTBF/T	exp(-T/MTBF)	Exact Av.	Approx. Av	Difference
0.16	6.25	0.852143789	0.924101319	0.92607189	1.97E-03
0.17	5.88	0.843664817	0.919618726	0.92183241	2.21E-03
0.18	5.56	0.835270211	0.915165492	0.91763511	2.47E-03
0.19	5.26	0.826959134	0.9107414	0.91347957	2.74E-03
0.20	5.00	0.818730753	0.906346235	0.90936538	3.02E-03
0.22	4.55	0.802518798	0.897641827	0.9012594	3.62E-03
0.25	4.00	0.778800783	0.884796868	0.88940039	4.60E-03

Table 3.4 Comparison of exact vs. approximate mean availability.

If the test interval is more than 20% of the MTBF, this approximation is not applicable. In such cases we can use a numerical solution such as the Maximum Likelihood Estimation technique—refer to Edwards[4] for details.

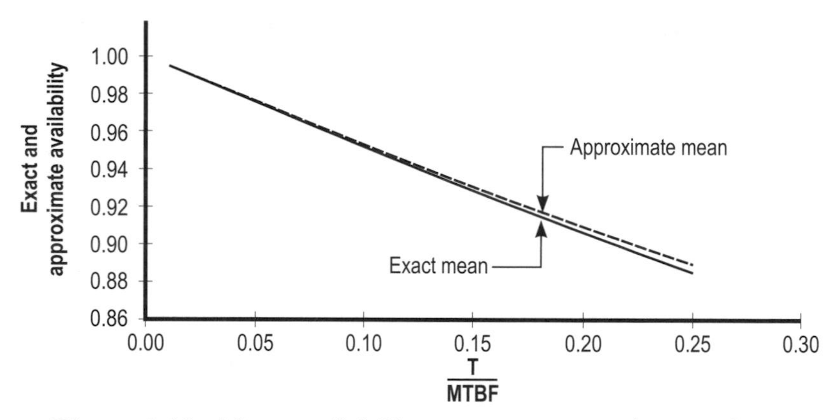

Figure 3.13 Mean availability; exact vs. approximate values.

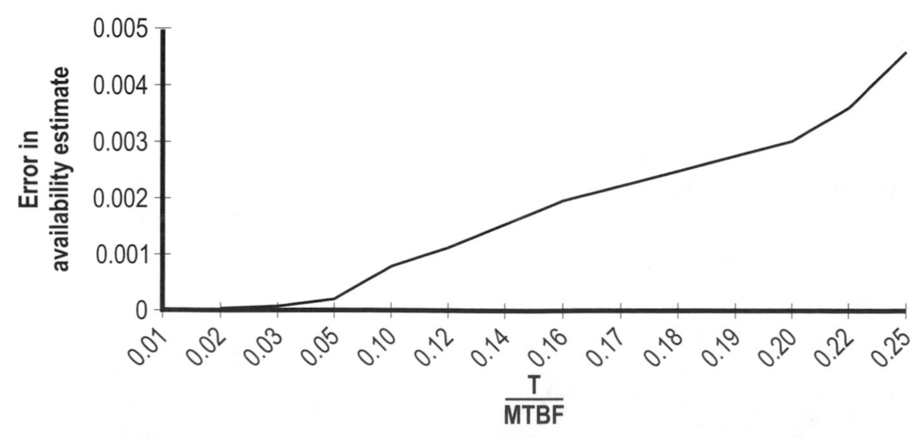

Figure 3.14 Error in estimate of availability vs. T/MTBF.

3.9 THE WEIBULL DISTRIBUTION

A number of failure distribution models are available.Among these are the exponential, gamma, pareto, Weibull, normal or Gaussian, lognormal, Birbaum-Saunders, inverse Gaussian, and extreme value distributions. Further details about these distributions are available in Hoyland and Rausand[3] or other texts on reliability theory.

Weibull[5] published a generalized equation to describe lifetime distributions in 1951. The two-parameter version of the Weibull equation is simpler and is suitable for many applications. The three-parameter version of the equation is suitable for situations where there is a clear threshold period before commencement of deterioration. By selecting suitable values of these parameters, the equation can represent a number of different failure distributions. Readers can refer to Davidson[6] for details on the actual procedure to follow in doing the analysis.

The Weibull distribution is of special interest because it is very flexible and seems to describe many physical failure modes. It lends itself to graphical analysis, and the data required is usually available in most history records. We can obtain the survival probability at different ages directly from the analysis chart. We can also use software to analyze the data.Figure 3. 15 shows a Weibull plot made using a commercial software application.

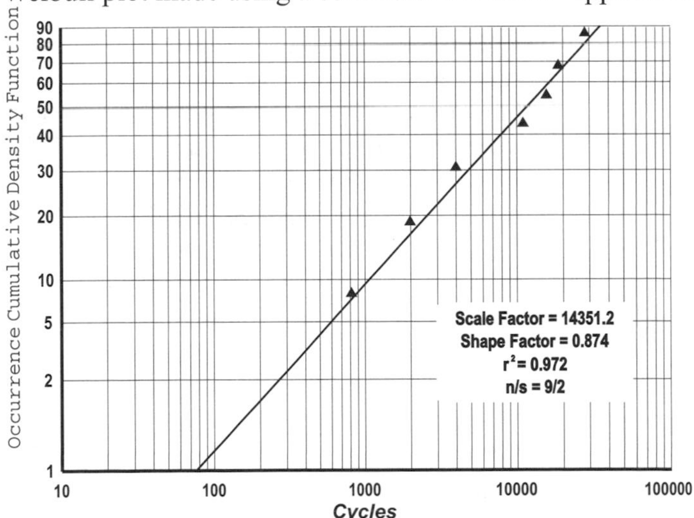

Figure 3.15 Typical Weibull Plot.

It is fairly easy to gather data required to carry out Weibull analysis, since time-to-failure and preventive replacement details for the failure mode are nearly all that we need. For this we need a record of the date and description of each failure. We also need the date of any preventive maintenance action that results in the component being repaired or replaced before failure occurs. Once we compute the values of the two parameters, we can obtain the distribution of failures. We can read the survival probabilities at the required age directly from the chart. This means that we can estimate the reliability parameters, and use this data for predicting the performance of the item.

The Weibull equation itself looks somewhat formidable. Using the simpler two-parameter version, the survival probability is given by the following expression.

$$R(t) = e^{-\left\{\frac{t}{\eta}\right\}^{\beta}} \qquad 3.14$$

where, η is called a scale parameter or characteristic life, and β is called the shape parameter.

Using expression 3.14, when $\mathbf{t} = \eta$, there is a 63.2% probability that the component has failed. This may help us in attributing a physical meaning to the scale parameter, namely that nearly 2/3rd of the items in the sample have failed by this time. The value gives us an idea of the longevity of the item. The shape factor β tells us about the distribution of the failures. Using expression 3.14, we can compute the R(t) or survival probability, for a given set of values of η and β at any point in time 't'. In appendix 3-2, we have provided the results of such a calculation as an example.

In spite of the apparent complexity of the equation, the method itself is fairly easy to use. We need to track the run-lengths of equipment, and to record the failures and failure modes. Recording of preventive repair or replacement of components before the end of their useful life is not too demanding. These, along with the time of occurrence (or, if more appropriate, the number of cycles or starts), are adequate for Weibull (or other) analysis. We can obtain such data from the operating records and maintenance management systems.

Such analysis is carried out at the failure modes level. For example, we can look at the failures of a compressor seal or bearing. *We need five (or more) failure points to do Weibull analysis.* In other words, if we wished to carry out a Weibull analysis on the compressor seal, we should allow it to fail at least five times! This may not be acceptable in practice, because such failures can be costly, unsafe, and environmentally unacceptable. Usually, we will do all we can to prevent failures of critical equipment. This means that we cannot collect enough failure data to improve the preventive maintenance plan and thus improve their performance. On items that do not matter a great deal—for example light bulbs, vee-belts, or guards—we can obtain a lot of failure data. However, these are not as interesting as compressor seals. This apparent contradiction or conundrum was first stated by Resnikoff[7].

3.10 DETERMINISTIC AND PROBABILISTIC DISTRIBUTIONS

Information about the distribution of time to failures helps us to predict failures. The value of the Weibull shape parameter β can help determine the sharpness of the curve. When β is 3.44, the ***pdf*** curve approaches the normal or gaussian distribution.High β values, typically over 5, indicate a peaky shape with a narrow spread. At very high values of β, the curve is almost a vertical line, and therefore very deterministic. In these cases, we can be fairly certain that the failure will occur at or close to the η value. Figure 3.16 below shows a set of ***pdf*** curves with the same η value of 66.7 weeks we used earlier, and different β values. Figure 3.17 shows the corresponding survival probability or reliability curves. From the latter, we can see that when β is 5, till the 26th week, the reliability is 99%.

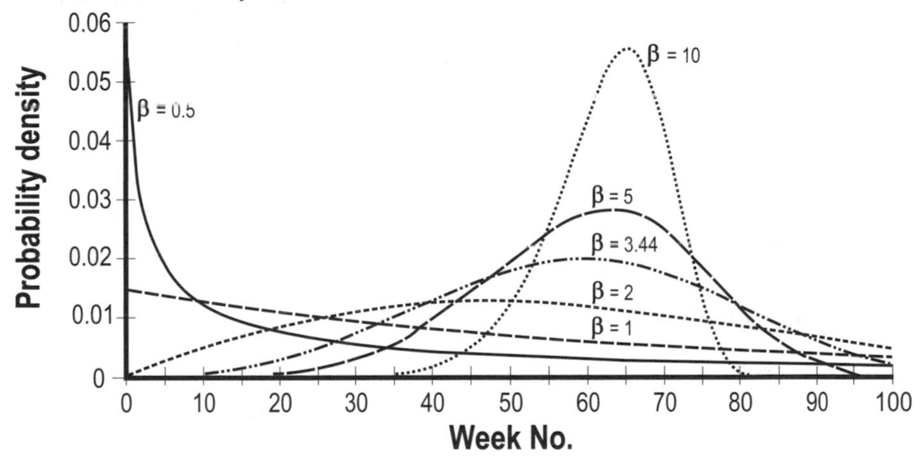

Figure 3.16 Probability density functions for varying beta values

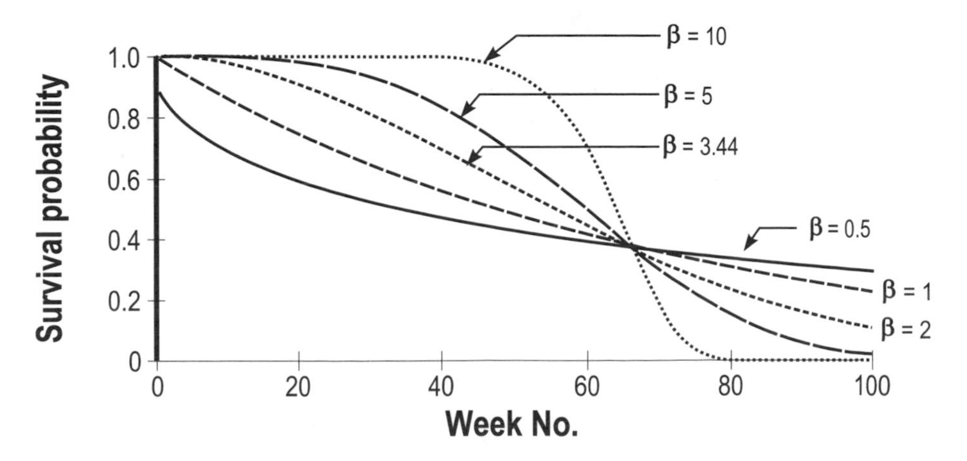

Figure 3.17 Survival probability for varying beta values

On the other hand, when we can be fairly sure about the time of failure, that is, with high Weibull β values, time-based strategies can be effective. If the failure distribution is exponential, it is difficult to predict the failures using this information alone, and we need additional clues. If the failures are evident and we can monitor them by measuring some deviation in performance, such as vibration levels, condition based strategies are effective and will usually be cost-effective as well.

If the failures are unrevealed or hidden, a failure-finding strategy will be effective and is likely to be cost-effective. Using a simplifying assumption that the failure distribution is exponential, we can use expression 3.13 to determine the test interval. *In the case of failure modes with high safety consequence, we can use a pre-emptive overhaul or replacement strategy, or design the problem out altogether.*

When β values are less than 1, this indicates premature or early failures. In such cases, the hazard rate falls with age, and exhibits the so-called infant mortality symptom. Assuming that the item has survived so far, the probability of failure will be lower tomorrow than it is today. Unless the item has already failed, it is better to leave it in service, and age-based preventive maintenance will not improve its performance. We must address the underlying quality problems before considering any additional maintenance effort. In most cases, a root cause analysis can help identify the focus area.

3.11 AGE-EXPLORATION

Sometimes it is difficult to assess the reliability of the equipment either because we do not have operating experience, as in the case of new designs, or because data is not available. In such cases, initially we estimate the reliability value based on the performance of similar equipment used elsewhere, vendor data, or engineering judgement. We choose a test interval that we judge as being satisfactory based on this estimate. At this stage, it is advisable to choose a more stringent or conservative interval. If the selected test interval reveals zero or negligible number of failures, we can increase it in small steps. In order to use this method, we have to keep a good record of the results of tests. It is a trial and error method, and is applicable when we do not have access to historical data. This method is called age-exploration.

3.12 CHAPTER SUMMARY

In order to evaluate quantitative risks, we need to estimate the probability as well as the consequence of failures. Reliability engineering deals with the methods used to evaluate the probability of occurrence.

We began with failure histograms and probability density curves. In this process we developed the calculation methodology with respect to survival probability and hazard rates, using numerical examples. Constant hazard rates are a special case and we examined their implications. Thereafter we derived a simple method to compute the test intervals in the case of constant hazard rates, quantifying the errors introduced by using the approximation.

Reliability analysis can be carried out graphically or using suitable software using data held in the maintenance records. The Weibull distribution appears to fit a wide range of failures and is suitable for many maintenance applications. The Weibull shape factor and scale factors are useful in identifying appropriate maintenance strategies.

We discussed age-exploration, and how we can use it to determine test intervals when we are short of historical performance data.

References

1) Nowlan F.S., and H.F.Heap. 1978. *Reliability-Centered Maintenance*. U.S Department of Defense. Unclassified, MDA 903-75-C-0349.
2) Knezevic J. 1996. Inherent and Avoidable Ambiguities in Data Analysis. UK: SRD Association, copyright held by AEA Technology, PLC. 31-39
3) Hoyland A., and M.Rausand. 1994. *System Reliability Theory*. John Wiley and Sons, Inc. ISBN: 0471593974. 18-72
4) Edwards, A.W.F. 1992. *Likelihood*. Johns Hopkins University Press. ISBN 0801844452
5) Weibull W. 1951. "A Statistical Distribution of Wide Applicability." Journal of Applied Mechanics, 18: 293-297.
6) Davidson J. 1994. *The Reliability of Mechanical Systems*. Mechanical Engineering Publications, Ltd. ISBN 0852988818. 22-33
7) Resnikoff H.L. 1978. Mathematical Aspects of Reliability Centered Maintenance. Dolby Access Press.

Appendix 3-1

DEVELOPMENT OF FAILURE PATTERNS

In order to understand these patterns, we will go through the calculation routine, using a set of artificially created failure data. We will use simplified and idealized circumstances in the following discussion.

In a hypothetical chemical process plant, imagine that there are 1000 bearings of the same make, type and size in use. Further, let us say that they operate in identical service conditions. In the same manner, there are 1000 impellers, 1000 pressure switches, 1000 orifice plates, etc., each set of items operating in identical service. Assume that we are in a position to track their performance against operating age. The installation and commissioning dates are also identical.

In table 3-1.1,* we can see the number of items that fail every week. We will examine six such elements, labeled A-F. The record shows failures of the originally installed items, over a hundred week period. If an item fails in service, we do not record the history of the replacement item.

Figures 3-1.1 to 3-1.6 illustrate the failures. If we divide the number of failures by the sample size and plot these along the Y axis, the resulting ***pdf*** curves will look identical to this set.

In each case, at the start there were 1000 items in the sample. We can therefore work out the number of survivors at the end of each week. We simply deduct the number of failures in that week, from the survivors at the beginning of the week. Table 3-1.2* shows the number of survivors.

Figures 3-1.7 to 3-1.12 are survival plots for the six samples.

We calculate the hazard rate by dividing the failures in any week, by the number of survivors.These are in table 3-1.3* and the corresponding hazard rate plots are in Figures 3-1.13 to 3-1.18.

These charts illustrate how we derive the failures, survivors and hazard rates from the raw data. As explained earlier, the data is hypothetical, and created to illustrate the shape of the reliability curves which one may expect to see with real failure history data.

*Note: In tables 3-1.1, 3-1.2, and 3-1.3, we have shown only a part of the data set. The data for weeks 11-44 and 55-90 have been omitted to improve readability.

Week No.	A	B	C	D	E	F
1	100	10	1	12	15	100
2	72	10	2	20	15	72
3	54	10	3	26	15	54
4	43	10	4	29	14	43
5	35	10	5	31	14	35
6	29	10	6	32	14	29
7	23	9	7	32	14	23
8	19	9	8	32	13	19
9	16	9	9	31	13	16
10	13	9	10	30	13	13
45	4	6	16	7	8	4
46	4	6	16	7	8	4
47	4	6	16	7	8	4
48	4	6	15	6	7	4
49	4	6	15	6	7	4
50	4	6	14	6	7	4
51	4	6	14	6	7	4
52	4	6	13	5	7	4
53	4	6	13	5	7	4
54	4	6	13	5	7	4
91	5	7	1	1	4	3
92	5	8	1	1	4	3
93	6	10	1	1	4	3
94	7	11	1	1	4	3
95	8	13	1	1	4	3
96	9	14	1	1	4	2
97	11	16	1	1	4	2
98	11	18	1	1	3	2
99	13	20	1	1	3	2
100	15	23	1	1	3	2

Table 3-1.1 Number of failures recorded per week—elements A to F

Note: Data for weeks 11-44 and 55-90 not shown

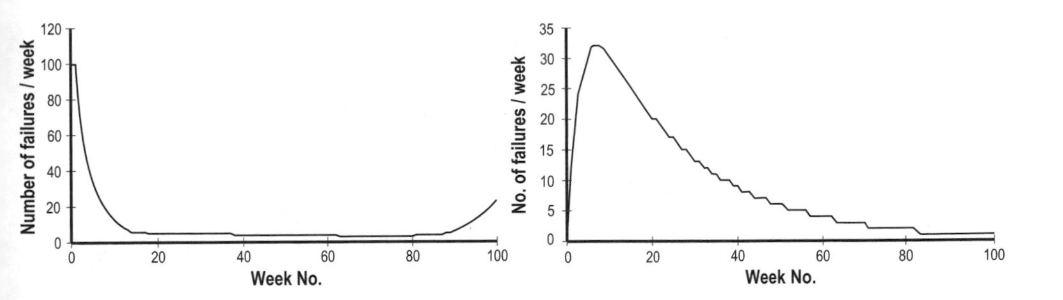

Figure 3-1.1
Failures recorded—element A.

Figure 3-1.4
Failures recorded—element D.

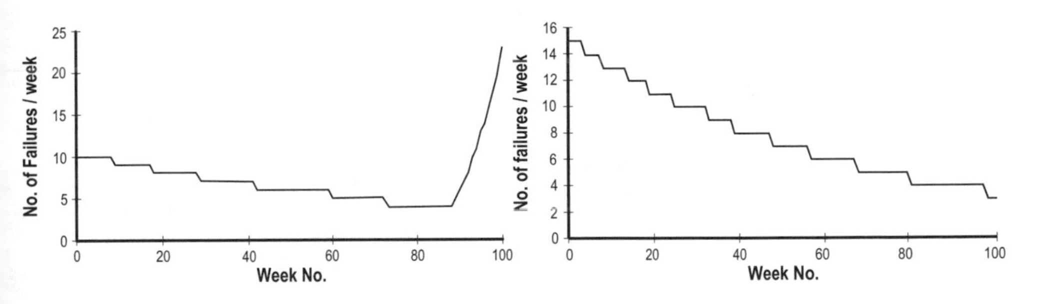

Figure 3-1.2
Failures recorded—element B.

Figure 3-1.5
Failures recorded—element E.

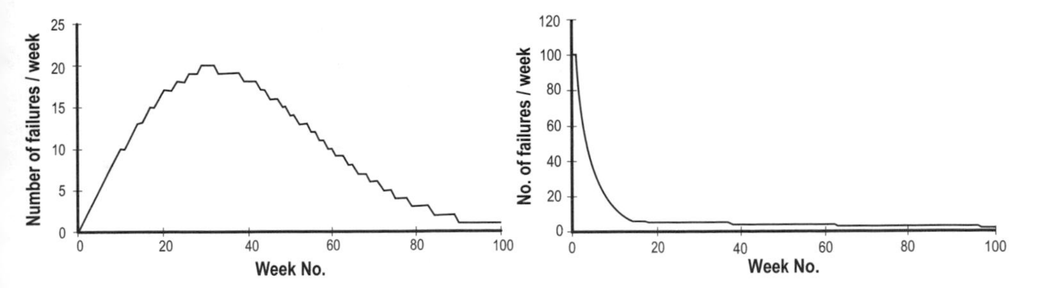

Figure 3-1.3
Failures recorded—element C.

Figure 3-1.6 Failures recorded—element F.

Week No.	A	B	C	D	E	F
1	900	990	999	988	985	900
2	828	980	997	968	970	828
3	774	970	994	942	956	774
4	732	961	990	913	941	732
5	696	951	985	882	927	696
6	668	941	979	850	913	668
7	645	932	972	818	900	645
8	625	923	965	787	886	625
9	610	914	958	755	873	610
10	497	904	946	725	860	597
45	413	636	350	168	507	413
46	409	630	333	161	499	409
47	404	624	318	154	491	404
48	400	617	303	148	484	400
49	396	611	288	142	477	396
50	392	605	273	136	470	392
51	388	599	259	130	463	388
52	385	593	246	125	456	385
53	381	587	233	120	449	381
54	377	581	220	115	442	377
91	255	394	13	24	253	260
92	250	386	12	23	249	257
93	244	376	11	22	245	255
94	237	365	10	22	242	252
95	228	352	9	21	238	250
96	219	337	8	20	234	247
97	208	321	7	19	231	245
98	197	304	7	18	227	242
99	184	284	6	17	224	240
100	169	261	5	17	221	237

Table 3-1.2 Number of survivors - elements A to F

Note: Data for weeks 11-44 and 55-90 not shown

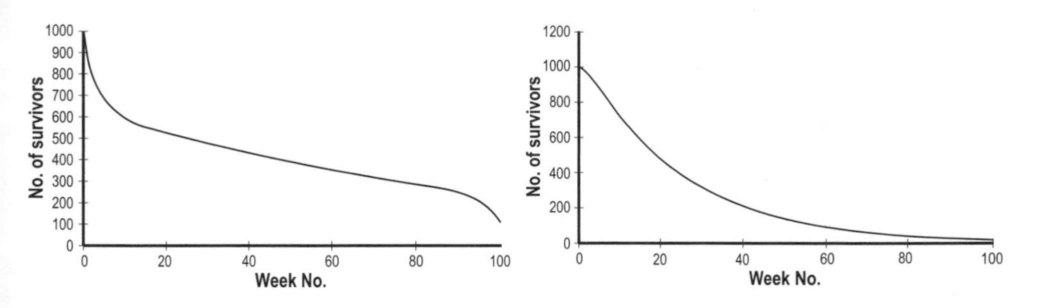

Figure 3-1.7 Survivors from original sample—element A.

Figure 3-1.10 Survivors from original sample—element D.

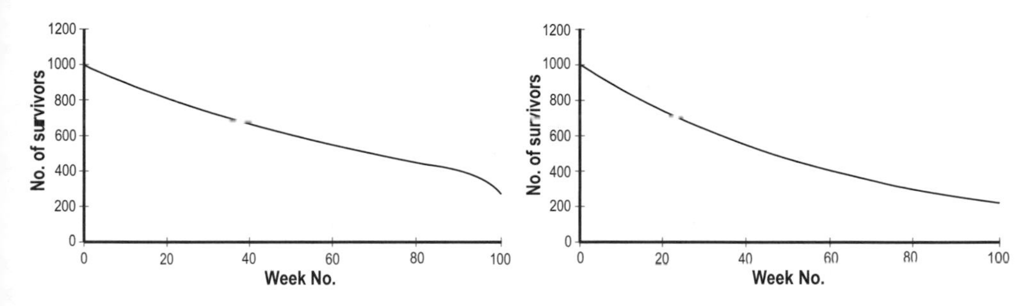

Figure 3-1.8 Survivors from original sample—element B.

Figure 3-1.11 Survivors from original sample—element E.

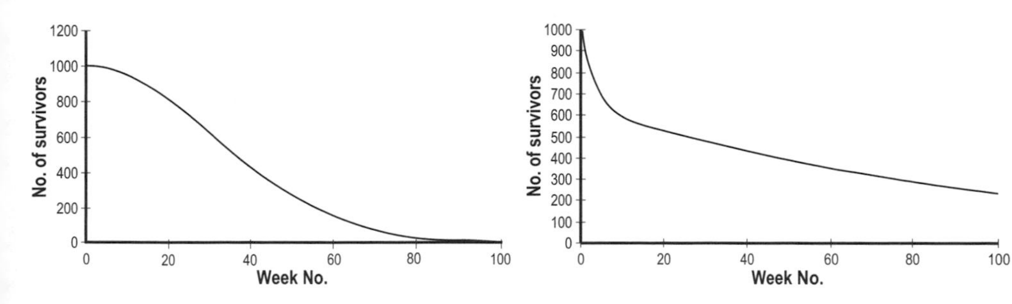

Figure 3-1.9 Survivors from original sample—element C.

Figure 3-1.12 Survivors from original sample—element F.

Week No.	A	B	C	D	E	F
1	0.1	0.01	0.001	0.012	0.015	0.1
2	0.08	0.01	0.002	0.02052	0.015	0.08
3	0.065	0.01	0.003	0.0265692	0.015	0.065
4	0.055	0.01	0.004	0.030864132	0.015	0.055
5	0.048	0.01	0.005	0.033903534	0.015	0.048
6	0.041	0.01	0.006	0.036078609	0.015	0.041
7	0.035	0.01	0.007	0.037615812	0.015	0.035
8	0.03	0.01	0.008	0.038707227	0.015	0.03
9	0.025	0.01	0.009	0.039482131	0.015	0.025
10	0.021	0.01	0.01	0.040032313	0.015	0.021
45	0.01	0.01	0.045	0.041	0.015	0.01
46	0.01	0.01	0.046	0.041	0.015	0.01
47	0.01	0.01	0.047	0.041	0.015	0.01
48	0.01	0.01	0.048	0.041	0.015	0.01
49	0.01	0.01	0.049	0.041	0.015	0.01
50	0.01	0.01	0.05	0.041	0.015	0.01
51	0.01	0.01	0.051	0.041	0.015	0.01
52	0.01	0.01	0.052	0.041	0.015	0.01
53	0.01	0.01	0.053	0.041	0.015	0.01
54	0.01	0.01	0.054	0.041	0.015	0.01
91	0.018	0.018	0.091	0.041	0.015	0.01
92	0.021	0.021	0.092	0.041	0.015	0.01
93	0.025	0.025	0.093	0.041	0.015	0.01
94	0.03	0.03	0.094	0.041	0.015	0.01
95	0.035	0.035	0.095	0.041	0.015	0.01
96	0.041	0.041	0.096	0.041	0.015	0.01
97	0.048	0.048	0.097	0.041	0.015	0.01
98	0.055	0.055	0.098	0.041	0.015	0.01
99	0.065	0.065	0.099	0.041	0.015	0.01
100	0.08	0.08	0.1	0.041	0.015	0.01

Table 3-1.3 Hazard rates - elements A to F

Note: Data for weeks 11-44 and 55-90 not shown

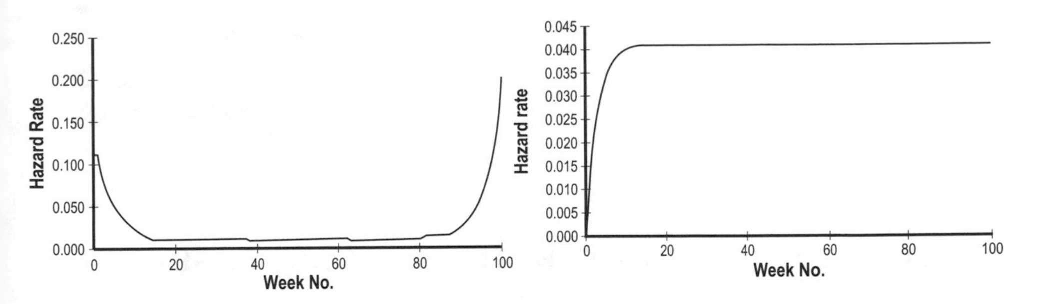

Figure 3-1.13 Hazard rate—element A. **Figure 3-1.16** Hazard rate—element D.

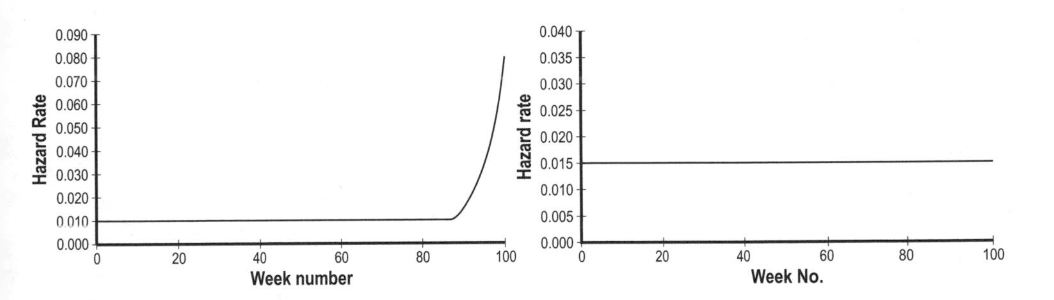

Figure 3-1.14 Hazard rate—element B. **Figure 3-1.17** Hazard rate—element E.

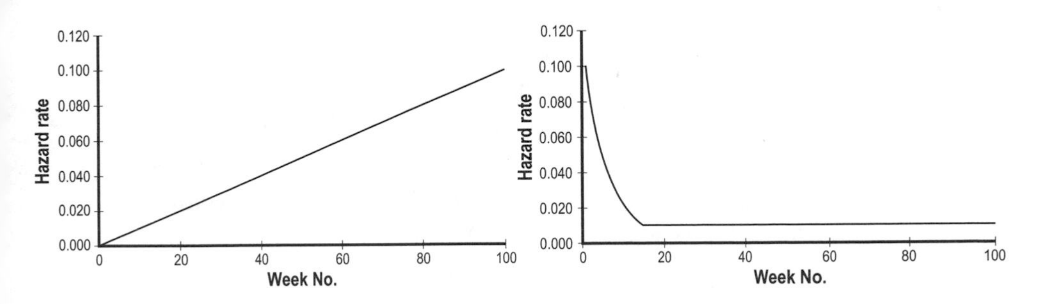

Figure 3-1.15 Hazard rate—element C. **Figure 3-2.18** Hazard rate—element F.

Appendix 3-2

AN EXAMPLE TO SHOW THE EFFECT OF THE THE SHAPE FACTOR

In Appendix 3-1, we derived the plots of the failure distribution, surviving population and hazard rates for a set of assumed data, to demonstrate the Airline Industry distribution of failures. In pattern E, namely, the constant hazard rate case, the value of the hazard rate is 0.015. In section 3.8 on mean availability, we discussed how the MTBF was the inverse of λ, which is the same as the hazard rate z(t) in the constant hazard case.

Thus the MTBF = 1/0.015 = 66.7 weeks. Recall that the MTBF is the same as the scale factor η, in the constant hazard case. So η= 66.7 weeks. We are now going to use this value of η, vary the time t, and use different values of β, and see how the distribution changes as β changes.

Using expression 3.14, we compute the R(t) for the data in Appendix 3-1, namely, η=66.7 weeks and for different values of β as t increases from 1 week to 100 weeks. From the R(t) value, we compute the cumulative failures F(t), which is = 1- R(t). The F(t) values are given below.

At low values of β, the distribution of failures is skewed to the left, i.e., there are many more failures initially than towards the end of life. In our example, at the end of 10 weeks, let us see how the β value affects F(t) up to that point.

When β=0.5, cumulative failures will be 32% of the total.
When β=1.0, cumulative failures will be 14% of the total.
When β=2.0, cumulative failures will be 2.2% of the total.
When β=3.5, cumulative failures will be <0.2% of the total.
When β=10, cumulative failures will be ~0% of the total, we do not expect any significantfailures till about the 32nd week.

Also of interest is what happens after we exceed the characteristic life. In week 77, i.e., ~ 10 weeks after the characteristic life is passed,

When β=0.5, cumulative failures will be 66% of the total.
When β=1.0, cumulative failures will be 68% of the total.
When β=2.0, cumulative failures will be 73% of the total.
When β=3.5, cumulative failures will be 80% of the total.

When $\beta=10$, cumulative failures will be 98% of the total.

From this, you can see that the higher the β value, the more the clustering of failures towards the characteristic value, and hence the greater predictability of time of failure.

At t=66.7 weeks, for all values of β, the R(t) is the same. In other words, the shape factor does not affect the survival probability when t = scale factor.

CHAPTER 4

Failure, Its Nature and Characteristics

In the last chapter we looked at aspects of reliability engineering that can be of use to the maintenance practitioner. We discussed some of the underlying principles that can help us identify reliability parameters from historical maintenance records. In order to apply this knowledge, it is useful to understand the nature of failure. In this chapter, we will examine the following.

- Failure in relation to the required performance standards; critical, degraded, and incipient failures;
- Significance of the operating context;
- Use of failures as a method of control of the process;
- Role of maintenance in restoration of desired performance;
- Incipiency and its use in condition-based maintenance;
- Age-related failures;
- System-level failures;
- Human errors and the effect of stress, sleep cycles, and shift patterns;
- Role of feelings and emotions and how these affect our reactions to situations.

4.1 FAILURE

4.1.1 Failure - a systems approach

Failure is the inability of an item of equipment, a sub-system, or system to meet a set of predetermined performance standards. This means that we have some expectations, which we can express quantitatively. For example, we can expect the discharge pressure of a centrifugal pump to be 10 bar gauge at 1000 liters per minute. In some cases, we can define our expectations within a band of acceptable performance. For example, the discharge flow of this pump should be 950-1000 liters per minute at 10 bar gauge. The performance standard may be for the system, sub-system, equipment, or component in question. These standards relate to what we need to achieve and our evaluation of the item's design capability and intrinsic reliability.

4.1.2 Critical and degraded failures

As a result of a failure, the system may be totally incapacitated such that there is a complete loss of function. For example, if a fire pump fails to start, it will result in the unavailability of water to fight fires. If there had been a real fire and only one fire pump installed, this failure could result in the destruction of the facility. In this case, the failure-to-start of the pump results in complete loss of function. As a second example, let us say that we have a set of three smoke detectors in an enclosed equipment-housing. The logic is such that an alarm will come on in the control room if any one of the three detectors senses smoke. If any two detectors sense smoke, the logic will activate the deluge system. It is possible that one, two, or all three detectors are defective, and are unable to detect smoke. When there is smoke, there is no effect if only one detector is defective, as the other two will activate the deluge. If two of them are in a failed state at the same time, the initiation of the deluge system will not take place when there is smoke in the housing. Lastly, with the loss of all three, even the alarm will not initiate. The loss of all three units will result in total loss of function, so this is a critical failure. If two of the three fail, the third can still initiate the alarm on demand. The operator then has the ability to respond to the alarm and initiate the deluge system manually. The system can still be of use in raising the alarm, so it has partial or degraded functionality.

4.1.3 Evident failures

When the impeller of a pump wears out, the operators can see the change in flow or pressure and hence knows about the deterioration in its performance. We call it an evident failure as the operator knows its condition. Similarly, an increase in the differential pressure across a filter or exchanger indicates an increase in fouling. When we take bearing vibration readings and plot the changes, it is possible to predict when it needs replacement. In each case, the operator knows the condition of the equipment, using their own senses or instruments. The operator, in this context, is the person who is responsible for starting, running, and stopping of the equipment. For example, the driver of an automobile is its operator.

4.1.4 Hidden failures

These failures, by contrast, are unknown to the operators during normal operation. Do you know if your automobile brake lights work? Similarly one does not know whether a smoke detector or a pressure relief valve is in a working condition at any point in time. A second event, such as a fire (causing smoke) will initiate the smoke detector, if it is in a working condition. If the vessel pressure exceeds the relief valve's set-pressure, it should lift. The standby power generator must start when there is a power failure. Will the pressure relief-valve lift or the standby generator start?

Hidden failures are also observed with protective instruments. Once equipment complexity increases, the designer provides various protective devices to warn the operator, using alarms, or bring it to a safe condition, using trips. These protective devices are rarely called upon to work and the operator will not know if they are working. These are subject to hidden failures.

If the operator is not physically present when the event takes place, is it an evident or hidden failure? For example, a pump seal may leak in a normally unattended unit. There will normally be some evidence of the leak, such as a pool of process liquid on the pump-bed. Merely because the operator was not present and did not see it does not change the event from an evident to a hidden failure. If the operator had been present, the leak would have been obvious, and a second event is not necessary. The question is not whether a witness was present, but whether the consequence occurred at the same time as the failure. To identify a hidden failure a second event must take place, and unless this condition occurs, it is an evident failure. Thus the time the operator sees the failure is not an issue.

To revert to the earlier question of the brake lights, you know that at the time you inspected the vehicle it was road-worthy, and the lights were working. If you ask a friend to stand behind the automobile while you press the brake pedal, you will soon know the answer. This is an example of a test on an item subject to hidden failures.

4.1.5 Incipient failures

If the deterioration process is gradual, and takes place over a period of time, there is a point where we can just notice the start of deterioration. Incipiency is the point at which the onset of failure becomes detectable. As the deterioration progresses, there is a point when the performance is no longer acceptable. This is the point of functional failure. The incipiency interval is the time from onset of incipiency to functional failure. When the failures are evident and exhibit incipiency, it is possible to predict the timing of functional failures.

4.2 THE OPERATING CONTEXT

The operating context describes the physical environment in which the equipment operates, demands made on it and the details of how it is used. The way in which we operate equipment has a bearing on how it performs, and affects its rate of deterioration. How close to the duty point does it operate? What is the external environment in which it operates? Does the internal environment affect its performance? What is the loading roughness? Does it have an installed spare unit that can come on stream if it fails? If the net positive suction head (NPSH) available to a pump is just acceptable, is the suction piping alignment such that the spare pump has as much NPSH as the duty pump? The answers to this type of questions will help define the operating context.

To illustrate this concept, let us take the example of an automobile or bus, and examine how we use it. For the purpose of this discussion, consider the following two contrasting requirements:

- We use it for long distance travel, mainly on freeways (highways, autobahns, or motor-ways);
- We use it for city travel only.

In the first case, the vehicle operates in a steady state, generally at cruising speed for much of its operating life. So the vehicle is operating predominantly at constant loads, well below duty point and with a smooth loading. In the second case, there will be frequent starts and stops, and driving speeds will be changing most of the time. The load on the engine will be variable due to the rapid changes in speed. The fluctuating power requirement from the engine means there will be more wear on the main elements of the power transmission, such as the clutch and gearbox. One should expect that brakes, tires, and indicator lights will need more frequent replacement.

We now add the driver profile, and the situation becomes more complex, for example,

- The driver has many years of experience, and has a 'clean' license, or
- The driver received the (first) license three weeks ago, and has already had one accident.

Turning to driving styles, we know that some drivers like to accelerate rapidly and use brakes frequently. Some are fond of taking corners at high speeds. Others prefer to cruise at a steady pace most of the time, use brakes infrequently, and take corners on all four wheels! Assume that you are buying a used car, and have the following options. One car belongs to a person who drives at a steady 40 mph, accelerates gently, and uses brakes sparingly. The other car, identical in make, model, vintage, and miles on the clock, belongs to a person who comes in screeching round the corner and slams on the brakes. If the price of the two cars is about the same, which one do you choose? It is an easy call, and you will decide quite quickly. The example highlights the significance of loading roughness, which contributes greatly to wear and tear.

External factors are next on our list of variables. These include dust or sand in the air, road surface, and weather conditions. One can see that the differences in performance as a result of these factors can be quite important.

Each of the changes in operating context will affect different sub-systems or components differently. For example, demanding driving habits will result in accelerated wear and tear on brakes, clutches, and tires, while dusty conditions will clog up air and lubrication filters more frequently. In an industrial context, the situation is quite similar. People wonder why identical pieces of equipment in the same process service perform differently. They believe that a pump is a pump is a pump! When we examine the differences in operating context, the reasons for performance variations become evident. As in the case of the vehicle, the operating context is one of the most significant contributors to performance.

4.3 THE FEEDBACK CONTROL MODEL

Let us examine how the driver of a vehicle controls it. The driver's eyes measure the position and attitude of the vehicle. These measurements are with respect to the edge of the road, other vehicles on the road, as well as any pedestrians who may be trying to cross the road. The change in position and attitude is being measured all the time. This information reaches the driver's brain, which compares these measurements with acceptance standards. The brain calculates the rate of change in position and attitude, and checks them against the norms. The driver's knowledge of the traffic regulations and past experience determine these acceptance criteria. The brain computes deviations from the norms, generating error signals. These signals initiate control actions, which are similar to those in section 1.4. The driver's brain instructs the hands to move the steering wheel, or the foot to press the brake or accelerator pedals, so that the car remains in control.

Other control systems follow a similar process, whether the unit in question is a battle-tank gun control or a chemical-process control system. Figure 4.1 shows a model illustrating the control mechanism.

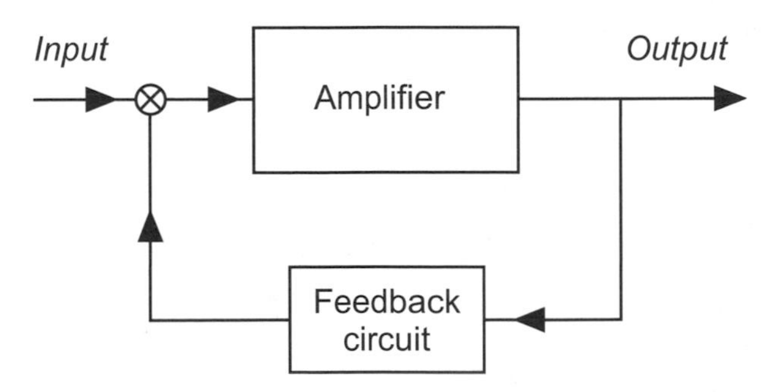

Figure 4.1 Input signal, amplifier, output signal, and feedback loop.

4.4 LIFE WITHOUT FAILURE

Would it not be wonderful to have life without failure? The fewer the failures, the higher the intrinsic reliability that we can enjoy. A good designer tries hard to make the product or service as reliable as possible, within given economic and technical constraints.

A marble rolling along a smooth glass surface may roll on for a long time. However, controlling its movement can be difficult. Similarly, an astronaut doing a space-walk faces a handicap. In the absence of friction or gravity, it is very difficult to navigate, because the only way to do so is to use reaction forces, applying the principle of conservation of momentum. Thus a lack of resistance or opposition may make the process energy-efficient, but control is more difficult. One could extend this approach to explain why democracies are superior to dictatorships, or why market forces are better than price controls. Seen in this context, failures can be useful, as they help identify deviations from expected performance and hence the scope for improvement.

Failures are deviations that we can measure, and provide the means to control a process. Resnikoff[1] identified the significance of failures when he presented his well-known conundrum. This is the fact that we require information about critical failures to identify the correct maintenance work, the purpose of which is to avoid the same failures. Hence with perfect maintenance, such critical failures will never take place, so we can never collect the relevant data! The inability to collect the data required for this purpose can stymie any organization attempting to go along the path of continuous improvement.

4.5 CAPABILITY AND EXPECTATION

Every component, equipment, or system has an intrinsic design capability. The bold line in Figure 4.2 shows this graphically.

The demand or expected value may be below this level, shown by the dashed line in Figure 4.2. In this case there should be no problem meeting the

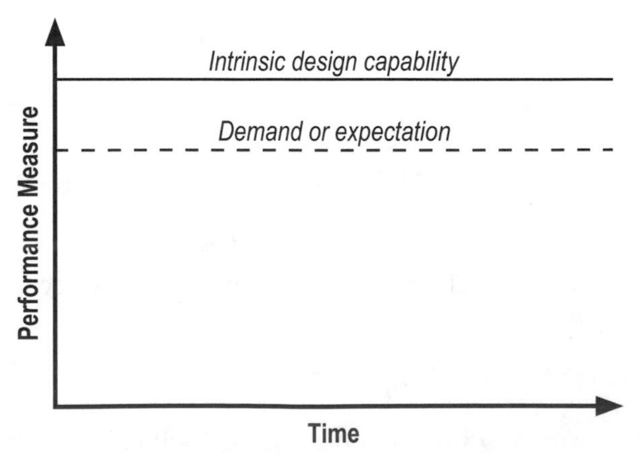

Figure 4.2 Normal relationship of demand to capability.

demand. However the expectation may be higher than the design capability, as shown by the dotted line in Figure 4.3. In this case, we cannot achieve the expected values on a long-term basis. No amount of maintenance can increase the capability of the equipment to produce continuously above the intrinsic design levels.

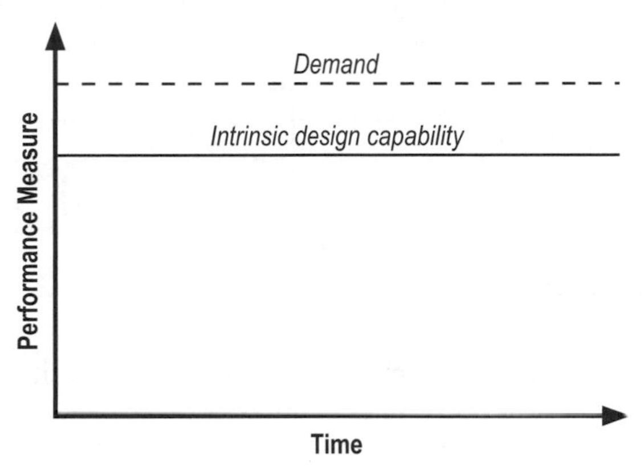

Figure 4.3 Demand exceeds capability.

Designers tend to build in some 'fat,' stating a level of capability lower than the real value. This is partly due to the use of standard components, some of which are stronger than required, and partly due to built-in safety factors. When we exploit this 'fat,' there is a temptation to think that we are able to exceed the design values continuously. The reality is that this capability was always there, but the designers informed us differently.

Over time, the capability line will droop, due to fouling, wear, fatigue, or chemical attack.When this happens, some maintenance has to be done, to bring the capability up to the design level, as shown in Figure 4.4.

The demand profile may be flat, or as is more common, fluctuating, with peaks and troughs. We cannot meet the expected demand when the two lines intersect, so we need to do some maintenance at this time. Alternatively, we can do the maintenance in anticipation of this situation as illustrated in Fig.4.5.

The capability line will also exhibit some roughness. Thus there will be a spread or distribution of values, in the case both of the capability line and the demand line. These can be shown as bands of values as shown in Figure 4.6 and its inset. Normally, with smooth demand and capability lines, there is a single point of failure, shown by point B in the inset. With both curves having a band of values the earliest point of intersection is point A and the latest point C. There is, therefore, a range of points of functional failure. This leads to uncertainty in determining it and the lowest value will normally be chosen, so that we are on the 'safe side.'

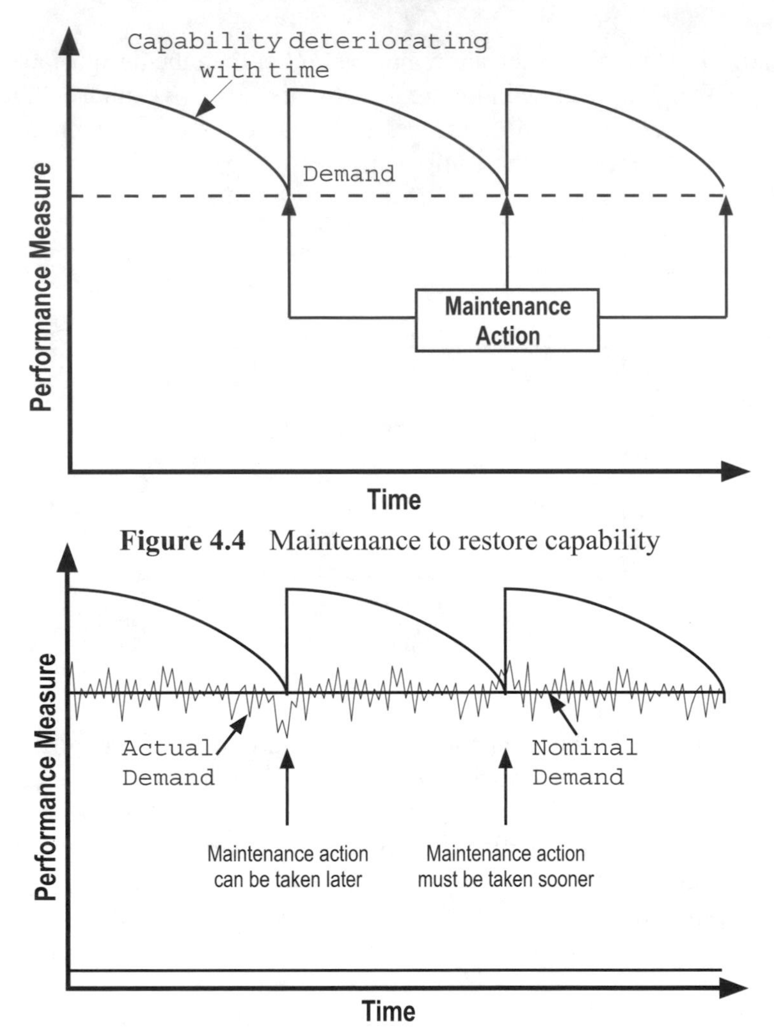

Figure 4.4 Maintenance to restore capability

Figure 4.5 Effect of demand fluctuations on maintenance timing.

4.6 INCIPIENCY

We mentioned incipiency briefly in section 4.1.5. Here we will examine the physical process in greater detail.

At the level of the smallest replaceable component, we will deal with items such as light bulbs, ball bearings, or structural welds. *Failure initiation is usually by fatigue or deformation caused by thermal or mechanical stress, or by chemical attack.* The rate of progression of the failure mechanism is variable, in some cases rapid, in others quite slow. Let us examine one or two common situations where we can observe the progress of the failure.

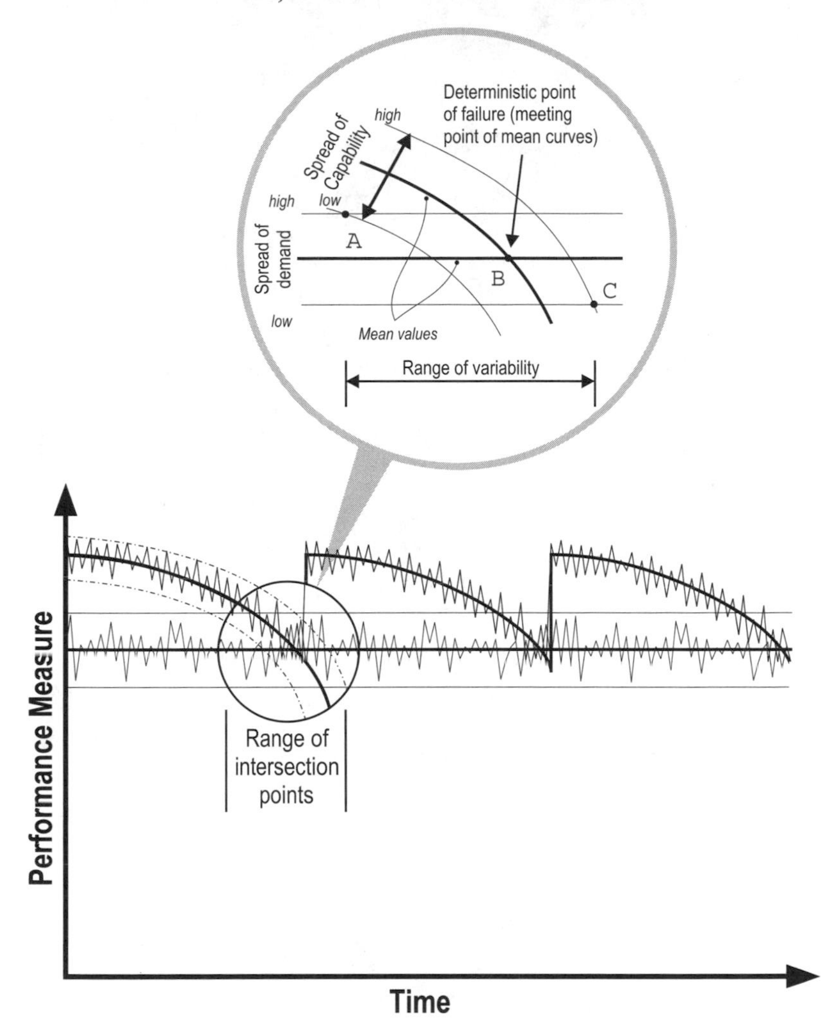

Figure 4.6 Effect of fluctuations in demand and capability on the timing of maintenance.

The first example is of a road that has a small surface defect or unevenness caused by poor finishing. As vehicles pass over this unevenness, the tires enter the depression and then climb up to the original level. This causes an impact load on the road as well as on the vehicle suspension. The effect of this impact on the road is to damage it further, causing a deeper depression. The next truck gets a bigger bump, and causes even more damage to the road. If we do not carry out repairs, the depression eventually becomes a pothole, making it unsafe to drive on this section of the road. Figures 4.7 to 4.9 below illustrate the sequence of events.

The time when we notice the initial defect is the start of the incipient failure, denoted by point x at time t_i in Fig.4.10 below. The droop of the curve shows the rate of growth of the pothole. At some point in time, this condition

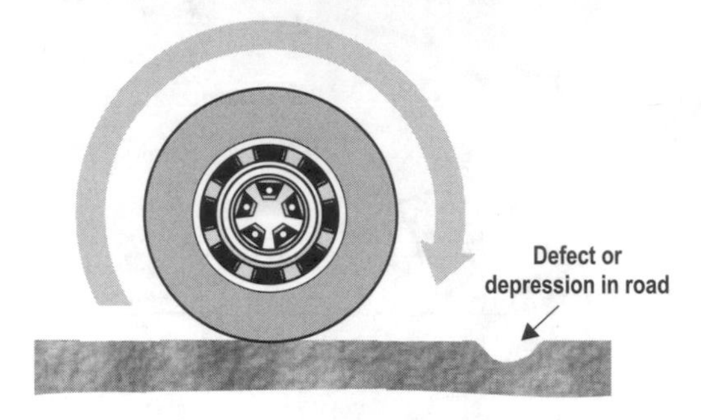

Figure 4.7 How road surfaces get damaged.

Figure 4.8 Tires 'drop' into defect and climb.

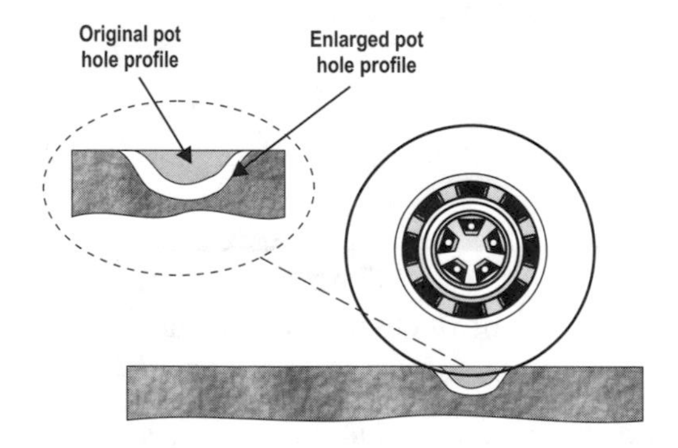

Figure 4.9 The 'drop' energy damages the road further.

becomes unacceptable, as the road is no longer safe to use. This norm used to determine its acceptability is dependent on the operating context. The higher the speed of the vehicles and the greater their loading, the stricter are the acceptance standards. The dotted lines show the relative levels of acceptability, which are dependent on road speeds and loading. At the point of intersection with the curve, indicated by the point y at time t_f, it is not safe to drive on the road any longer. In other words, it has failed. *The time taken for the condition to deteriorate from x to y, that is,* $t_f - t_i$, *is the incipiency interval.*

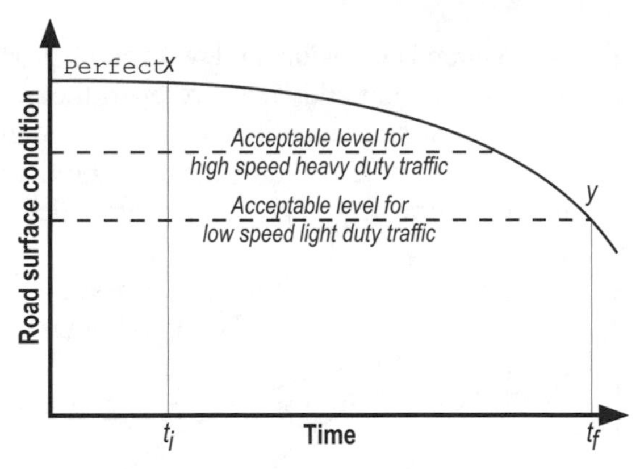

Figure 4.10 Incipiency interval $(t_f - t_i)$

The second example is of a welded structure, such as a pressure vessel or steel frame of a building. When originally fabricated, some minor cracks would have remained in the welds. At the time of construction, these cracks either escaped detection or were not serious enough to trace and repair. After commissioning the structure, these welds experience loads, which can fluctuate in magnitude, direction or both. When there are cracks in the welds, the effective cross sectional area is smaller, resulting in higher stresses. At the tip of the crack (refer to figure 4.11) the material can become plastic due to stress concentration.The most stressed part of the weld will yield, resulting in the crack propagating further. This raises the stress just beyond this point, ensuring the continuous propagation of the crack. In due course, the crack can grow to such an extent that the weld as a whole is no longer able to perform its function.

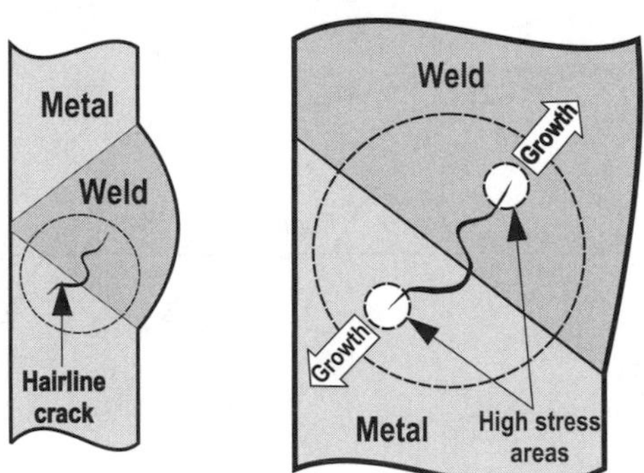

Figure 4.11 Crack propagation in a weld.

The incipiency interval may be very short, as in the case of light bulbs, or very long, as in the case of weld crack propagation. A large number of failures have incipiency intervals ranging from weeks to several months or years. Bearing failures, general corrosion, and weld crack propagation are all examples of such failures. Nowlan and Heap[2] refer to the point x in Figure 4.10 as the point of potential failure, and the point y as the point of functional failure. Moubray[7] refers to it as the P-F curve, where points P and F correspond to points x and y in Figure 4.10. The range of variance in incipiency is shown in figure 4.12.

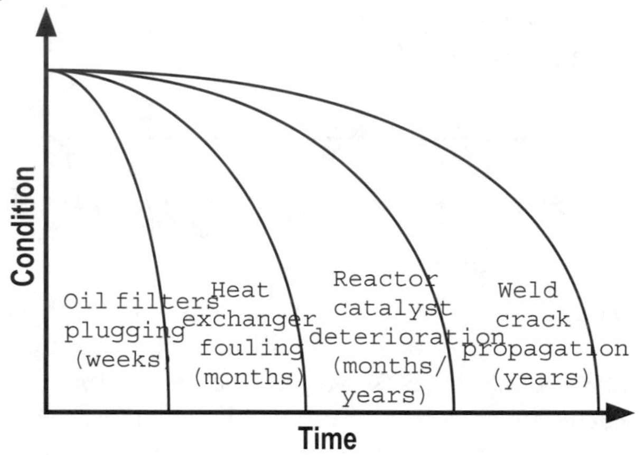

Figure 4.12 Examples of incipiency intervals.

Even in the case of a single failure mode in a given operating context, the droop of the incipiency curve may vary. Thus, there is a range of incipiency intervals, as illustrated in Fig.4.13. This range introduces uncertainty in determining the incipiency interval.

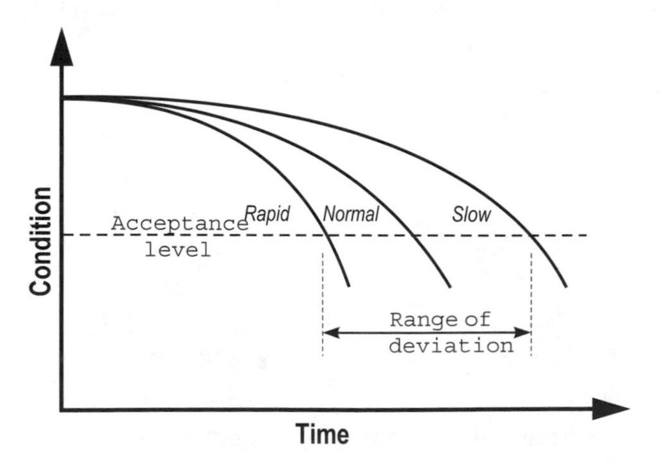

Figure 4.13 Variations in incipiency intervals.

4.7 LIMITS TO THE APPLICATION OF CONDITION MONITORING

When the incipiency is very short, the time available to plan or execute maintenance action is also very small. In such cases, it is difficult to plan replacement before failure by monitoring the component's condition. When incipiency intervals are in weeks, months, or years, condition monitoring is often an effective way to plan component replacement. Condition monitoring is feasible when it is possible to measure the change in performance, using human senses or instruments. It follows that we cannot monitor hidden or unrevealed failures.

Proponents of condition-based maintenance are correct when they highlight their ability to predict failures. Any predictive capability enhances the decision making process. However they sometimes give the impression that condition monitoring systems will solve all our problems. We know that all failures do not lend themselves to condition monitoring. The failure must exhibit incipiency, it must be feasible to measure it, and the interval must be of reasonable duration. We must always ask the providers of condition monitoring services to demonstrate how they meet these requirements.

4.8 AGE RELATED FAILURE DISTRIBUTION

A system consists of many pieces of equipment, each of which has several components. Each component can fail in one or more ways. In Chapter 3, we looked at the six failure patterns identified by the Nowlan and Heap[2] team. You will recall that these failure patterns are plots of the hazard rates against time. The author obtained similar results in a study of failures in the offshore oil and gas industry. The experience of other industries is not available in the public domain, but we can expect broadly similar results.

Prior to the Nowlan and Heap study, the belief was that all failures followed the so-called bath-tub curve. Their results showed that this pattern was only applicable to 4% of all the failure modes.

Fourteen percent showed a constant failure pattern, and if we ignore the failures that took place early in life, a further 75% also followed this pattern. The remaining 11% (including 4% of the bath tub) of the failure modes exhibited a distinct relationship to age. Should we concern ourselves with this relatively small proportion of failures that exhibit an age-relationship?

To answer this question, we need to know whether any of these failure modes could result in serious consequences. If so, they acquire a new level of respect. With a skewed distribution, a strategy based on an assumed constant failure pattern will not be satisfactory. *Therefore, we cannot assume that all failures exhibit a constant hazard rate pattern, as long as any of the remaining 11% matter.*

4.9 SYSTEM LEVEL FAILURES

When we assemble components to build equipment, each component failure-mode affects the overall failure rate. These individual component failure-modes may have exhibited a distinct age-related failure pattern. When any failure takes place, we replace the affected part with a new one. In an ideal case, we do not replace any of the other components at this point. The latter are at different stages of deterioration in their own life cycles. One of these will fail some time thereafter because it has reached the end of its life. We replace it and start a new cycle, while other components continue from their partly worn-out state.The result is that at the assembly level, the failures tend to follow the exponential distribution.

The concept of Mean Time To Failures, or MTTF, is worth further consideration at this point. As discussed in Chapter 3, the mean does not tell us much about the distribution. With a given sample, many of the failures could have taken place early or late in terms of age. In such a case, the use of the mean distorts the picture, because one may wrongly infer that the failures take place uniformly over the life. Hence, the use of MTTF without a full understanding of the distribution may lead to inappropriate decisions.

When the hazard rate is constant and the distribution is exponential, it is perfectly acceptable to use the MTTF. At this point there is (approximately) a 63% probability that the component has failed, and only a 37% probability of survival. In cases where the consequences of failure are high, we must do whatever we can to reduce or eliminate them. If the failure is evident and exhibits incipiency, for example, as in a ball bearing, we can take vibration or other condition monitoring action. If the failure is hidden, as for example, in a gas detector, we carry out a test, or a failure finding task. We must plan preventive maintenance action well before t = MTTF, because we cannot accept a 37% probability of survival at the time of the test or repair. The lay person often thinks of the MTTF as the expected time of failure and, therefore, the maintenance interval, which is clearly not the case.

4.10 HUMAN FAILURES

Nearly three quarters of all accidents are due to the action (or inaction) of human beings. We cannot wish it away, as it is too large a contributor to ignore. Human beings are complex systems, with hundreds of failure modes. In the following discussion we will use the terms *human error* and *human failure* interchangeably.

The causes of human error are many and varied. Lorenzo[3] categorizes them as random, systematic, and sporadic. We can correct random errors by better training and supervision. A shift in performance in one direction indicates systematic variability. We can reduce these by providing a regular performance feedback. Sporadic errors are the most difficult ones to predict or control. In this case, the person's performance is fine for most of the time. A sudden distraction or loss of concentration results in sporadic error.

There is an optimum level of stress at which human beings perform well. *A certain level of stress is necessary to keep us alert, active, and expectant.* We call this facilitative stress. Too high a stress level can be as a result of physical or psychological pressures. This may result in tiredness and lack of concentration. Too low a stress can be due to the work being repetitive, intellectually undemanding, or otherwise boring. During World War II, the British Royal Navy noted that submarine lookouts became ineffective after about 30 minutes, as they could not remain alert, The lookouts knew that their own lives depended on their vigilance, so motivation was not an issue.

Swain and Guttmann[4] give the following examples of psychological stress:

- Suddenness of onset
- Duration of stress
- Task speed
- Task load
- High jeopardy risk
- Threats of failure, loss of job
- Monotonous, degrading, or meaningless work
- Conflicts of motives about job performance
- Reinforcement absent or negative
- Sensory deprivation
- Distractions such as noise, glare, flicker, color, or movement
- Inconsistent cueing

Each person is slightly different and thrives under different levels of stress. However, a number of the stress factors affect many people in similar ways.

In order to reduce human failures, we have to address the factors contributing to stress. By doing so, we can produce the right environment for each person. In most cases, we will not be able to influence stress caused by domestic matters, so we will focus on those at work. Job enrichment deals with the elimination of boredom and unacceptably low stress levels. We can attribute the remaining problems to high stress at work.

Control room operators perform critical functions. During plant upsets, startups, and shutdowns, their skills are in demand. We use alarms to catch their attention when things go wrong. Designers of control rooms have to take care to minimize the number of alarms they install. If too many alarms come on too quickly during a plant upset, operators can lose concentration and react incorrectly, thereby worsening the situation. In an article entitled 'How Alarming!,' Bransby and Jenkinson[5] report the results of a survey. They studied 96 control room operators in 13 different plants in the U.K. Their findings, listed below, indicate that we have to devote more attention to this issue at the design stage.

- In an average shift, during steady operations, operators receive an alarm about every two minutes;
- Many of these are repeats of ones that occurred in the previous five minutes;
- Operators stated that many of them were of little value to them, and that eliminating about 50% would have little or no effect;
- Following a plant disturbance, they estimated that there were about 90 alarms in the first minute and seventy in the next ten minutes;
- About half the operators said that they felt forced to accept alarms during plant upsets, without reading or understanding them;
- During the survey, they observed one such plant upset. The operator did not make a full check of the alarms for about half an hour. This behaviour was consistent with that reported by the others in the survey.

Since the purpose of the alarm is to alert the operator, these results indicate that the designers have failed in their objectives. The authors state that improvements are possible, and that a variety of tools are available. Some of the simpler ones include tuning-up limit values and dead-bands, and adjusting priorities. The use of logic to suppress some non-essential post-trip alarms is also possible. As an example, they state that a review of the alarms resulted in a 30% reduction in the number of alarms.

One of the causes of human failures is tiredness, and this is often due to sleep deprivation. The human body operates with the help of a biological clock. Shift work can disturb normal (or circadian) sleep cycles. As a result, the reaction to stimuli can be slow. This can affect the ability of the operator to respond to a rapidly developing scenario. Night shift workers are more susceptible to this problem than the rest, because of the disturbance to their circadian rhythm. While there is no direct cause and effect relationship established, we note that some of the worst industrial disasters including Piper Alpha, Bhopal, Chernobyl, Three-mile Island, and Exxon Valdez occurred in the silent hours. This does not automatically mean that it is unsafe to work at night. Night-shift workers have completed many millions of hours of work without any incidents. It is the combination of circumstances that matter, so one must view this in context. Since we cannot eliminate night shift work, especially in continuous process plants, we have to try to understand the risks, so that we can take suitable steps to minimize them.

A factor affecting sleep cycles is the way we arrange shift patterns. Lardner and Miles[6] have explained why some shift patterns are superior to others from an ergonomic point of view. They propose a nine-day cycle, with 2 days each in the morning, afternoon, and night shifts, with a

3-day 'weekend' following the night shift. The 'weekend' may turn out to be in the middle of the week. They argue that this pattern is superior to the alternative 28-day cycle, which is quite common. The 28-day cycle consists of 7 night shifts and 7 evening shifts, followed by a 2 day 'weekend' after each block. This is followed by 7 morning shifts and a 3 day 'weekend.'

Human errors occur due to a number of reasons, and lack of knowledge and experience are not necessarily the most common. Motivation and morale are often key issues to manage. Pride in work, a sense of being wanted, and being treated fairly are all important considerations. We all want 'user friendly' software; similarly, staff want managers who are 'people friendly.' When this is so, we are likely to experience lower absenteeism or sickness, better participation in team effort and suggestion schemes, lower accident rates, and higher productivity.

What makes human beings distinctly different from machines is their ability to think, often in a very creative manner. Feelings and emotions change the way a person responds to identical stimuli over time, and makes it hard to predict behavior. We have provided a brief introduction to the subject in this chapter and readers can refer to Lorenzo's excellent guide for a more detailed discussion. A check-list of potential causes of human errors is available in Appendix 4-1.

4.11 CHAPTER SUMMARY

We began this chapter by defining failure in relation to the required performance standards. Failures can be critical, causing total loss of function, degraded where the loss is partial, or incipient where progressive deterioration has commenced, but will take some time before there is loss of function. We note the significance of the operating context, and how this explains why identical items of equipment perform differently. We saw how failures themselves provided a means of control on the process.

Our next topic is the role of maintenance in achieving the desired equipment performance, as long as it is lower than its capability. We discussed incipiency, and its use in condition-based maintenance, using some common examples to illustrate the concepts. Thereafter, we discussed age-related failures.

Finally, we looked at human errors, perhaps the most complex issue relating to failures. We noted that there is an optimum level of stress required to keep human errors as low as possible. The work done by experts on sleep cycles shows us how they can affect the body's natural rhythm. The experts state that some shift patterns are superior to others when planning 24-hour coverage for continuous process plants.

Feelings and emotions play a major role in affecting the way people react to situations. Therefore, managers have to focus on motivation and morale, which are key issues in minimizing human failures.

References

1) Resnikoff, H.L. 1978. Mathematical Aspects of Reliability Centered Maintenance. Dolby Access Press.
2) Nowlan, F.S., and H.F. Heap. 1978. *Reliability-Centered Maintenance.* U.S. Department of Defense. Unclassified, MDA 903-75-C-0349.
3) Lorenzo, D.K. 2001. *A Manager's Guide to Reducing Human Errors: Improving Human Performance in the Process Industries.* API Publication 770.
4) Swain, A.D., and H.E. Guttmann. 1983. *Handbook of Human Reliability Analysis with Emphasis on Nuclear Power Plant Applications.* NUREG/CR-1278-F SAND80-200. 3-34.
5) Bransby, M., and J. Jenkinson. 1998. "How Alarming!" The Chemical Engineer. January 15: 16.
6) Lardner, R., and R. Miles. 1997. "Better Shift Systems." The Chemical Engineer. September 11: 28.
7) Moubray, J. 2001. *Reliability-Centered Maintenance*. Industrial Press, Inc. ISBN: 0831131462.

Appendix 4-1

ERROR PRONE SITUATIONS

Reproduced courtesy of the American Petroleum Institute (see Reference 3 above).

A check-list of work situations that could lead to human errors is listed below, based on Lorenzo.

- incomplete, inadequate, out of date, or non-existent procedures
- poor or misleading instrumentation
- lack of competence and knowledge
- conflicting priorities, especially between safety and production
- poor labelling
- inadequate feedback
- non-enforcement of policies and procedures
- excessive spurious trips, causing protective instruments to be defeated
- poor communications
- unsatisfactory plant layout
- control systems that are over-sensitive
- mental overload during emergencies
- error prone situations, typically with excessive manual operations, inadequate interlocks, or wrong use of interchangeable fittings
- improper tools and test equipment
- poor housekeeping
- excessive demand on operator vigilance
- software or control hardware faults
- poor ergonomics

CHAPTER 5

Life Cycle Aspects of Risks in Process Plants

Every process plant goes through its design, construction, commissioning, operating and decommissioning phases. In this book the term *process plant* covers any plant that uses the production or distribution process as defined in Chapter 1. It includes, for example, utility companies, paper and steel mills, and transport companies. As long as the product or service handled is physical, the principles are applicable to all of these plants. We can minimize the risks associated with each of these phases when we know the contributing causes. In this chapter, we will focus our attention on these life cycle risks, and cover the following areas:

- Quality of design and intrinsic reliability of the plant;
- Importance of simplicity in designs;
- Risks in the construction and commissioning phases;
- Design changes and the high level of associated risks; importance of change-management;
- Maintenance cost-drivers; risks associated with unstructured cost reductions; ways to reduce costs without losing control on safety and profitability;
- Process plant end-of-life activities and associated risks.

Commissioning a new plant can be an exhilarating or frustrating experience, depending on how well the designer has anticipated start-up problems, and whether the plant functions as required. It is not unusual to find a number of change requests being initiated during and shortly after commissioning the plant. If the change requests relate to the original functional requirements, operability, or maintainability, they indicate deficiencies in the design.The number of such requests is one measure of the level of dissatisfaction.

Other change requests relate to the desire to increase plant capacity. By operating new plants at design and higher-than-design throughputs, we can test them. Some equipment, piping, or logistics will stand out as bottle-necks. Lack of balance between the different parts of the plant is the cause of these bottle-necks. Change requests that relate to the removal of these bottle-necks are capacity-increase projects. This type of de-bottlenecking could lead to reliability problems.

We cannot avoid some of this imbalance, for which there are several contributing factors. First, when the designer needs items such as a length of pipe, a centrifugal pump or a gas turbine, the vendors would offer it in a standard range of sizes. The designer does not have the choice of trimming the sizes. As long as the item on offer is close to the specifications and budget, it is acceptable. Hence the selected items are usually larger or stronger than required. Second, there is always a residual amount of uncertainty in any new design, in spite of all the analysis and expert inputs. The designer will build in some 'fat' to take care of these uncertainties. Third, there may be bonus or penalty clauses in the contracts to ensure that the plant design meets its functional requirements. Turnkey contracts often have such provisions. The cost of building in a little extra capacity is usually quite small in comparison to these bonuses and penalties. The designer avoids the penalties and adverse publicity by building in some over-capacity. Last, the designer uses redundancy to guarantee the reliability of the plant. Sometimes installed spare equipment is necessary for safe and reliable operation of the plant. However, in many cases, custom and practice dictate the decision-making process. The correct method is to carry out a risk analysis before choosing installed spare capacity. However company standards and codes of practice often mandate such practices. All of these factors contribute to over-capacity or fat in some parts of the plant.

We often purchase oversized equipment without realizing that this is happening. As an illustration, consider the selection of a centrifugal pump. The sequence of events is often along the following lines:

1. The process designer calculates the discharge pressure required to overcome the back pressure at the rated flow, the available suction head, and the drop in the piping, valves, and fittings. This includes an allowance for uncertainty.
2. The instrument designer adds the pressure drop across orifice plates and control valves, again including an element for uncertainty.
3. The project engineer writes the requisition for the pump, and invites bids from vendors.
4. The buyer's equipment specialist looks at the pump selection charts among the offers received, and selects a suitable pump, usually the next size above the required capacity. The selection charts show the flow and pressure combinations that a given model can provide.
5. In producing the selection charts, the vendor has allowed for some manufacturing deviations and de-rated the equipment slightly. This gives the vendor a comfort cushion to cater for uncertainty.
6. As a result of all these allowances, the pump discharge pressure can be, say, 20-40% higher than required at rated flow.
7. This additional pressure energy will be dissipated as heat, vibration, and noise in the control valve.

Admittedly, there is some exaggeration in this example, but it is not far off the mark. If you take a walk in a chemical plant or petroleum refinery, you are likely to find some noisy control valves on pump discharge lines. The valve body can be quite hot, and may even have blistered paint-work. Further examination will reveal that the pump's discharge pressure is excessive, and that this additional energy is being dissipated in the control valve. Apart from the fact that energy is being wasted, the pump is also operating with a throttled discharge. This causes excessive wear, as the internal leakage past the wear rings will increase. The local flow rate inside the control valve can be very high, resulting in erosion of the trim. The probability of failure increases, both of the pump and the control valve. Due to the additional erosion inside the bodies, the physical damage to the internals is larger than otherwise. Thus the consequence of failure is also higher, as repair costs rise. Depending on the level of redundancy built-in, the loss of the pump could result in an immediate operational consequence.

The risk of failure in such cases is, therefore, considerably higher than with a less conservative design. Unreliability and over-capacity are built in due to these provisions for uncertainty. In an extreme case, the higher failure rate can result in the system availability dropping to an unacceptable level, thereby defeating the design intent. Finally, the capital cost and the power consumption increase, so we end up losing on all fronts. This example illustrates the fact that conservative and 'safe' designs can result in increasing the risk to the owner.

The issue of over-capacity is not as simple as it may appear at first sight. It depends on external constraints, and the designer's skill. In most cases, over-capacity simply means additional capital and operating costs. It may also result in reduced overall reliability and availability, thereby reducing the plant's profitability.

5.1 DESIGN QUALITY

A well-designed plant will have some distinct features, which include the following:

- The plant is able to produce products of the desired quality consistently;
- The rate of production is satisfactory;
- The production process is efficient;
- The plant is easy to operate;
- The plant is easy to maintain;
- The plant is reliable.

The first three points describe the aptness or functionality of the plant. In other words, the plant is capable of producing the required output, with the designed inputs of materials, energy, and human effort. However, it will be safe and profitable only if it meets the remaining three conditions. The exposure to safety or environmental incidents is higher in plants that are difficult to operate. If this is so, operators will find work-around solutions to their problems. Their make-shift efforts can lead to unwanted incidents as they do not have training or experience in design. Similarly, repair times will be excessive in plants that are difficult to maintain. This results in low availability of protective devices and production equipment, thereby adversely affecting safety and profitability. Unreliable plants suffer from frequent trips or breakdowns, which result in production losses and additional work for the operators and maintainers.

It is reasonable to expect that designers will strive hard to meet these six requirements, but they will not necessarily succeed all the time. Let us therefore, examine why the design quality is less than optimal. These fall in one or more of the following categories:

- Insufficient information is available to the designer in respect of the required functionality;
- The design team is under severe resource and time pressure;
- The design team lacks the required knowledge, experience, and skills;
- The customer requirements have changed since the time the plant was conceived.

A poor design will result in a problem plant throughout its life. Once the plant is in operation, the maintenance manager will try to find solutions, but these will generally be short term, low-cost fixes. Only a permanent solution that addresses the root causes will eliminate this problem. It is important to get the design right the first time, as the alternative is a potentially unsafe or undersized plant, perennially in trouble. In order to do so, it is a good practice to involve the relevant people in the organization, right from the inception of the project. The marketing, operations, and maintenance staff can provide the relevant inputs.

5.1.1 Marketing inputs

The inputs from the marketing experts will help define the product volume, growth rate, and the customer expectations in respect of quality and functionality. In the case of some consumer-goods industries, the market may fluctuate considerably, making predictions of volume quite difficult. Competitors' actions also influence the market and, in some cases, even the functional requirements can change over relatively short periods. Thus at the time of commissioning, the design may not meet the new functional requirements. In these cases, flexibility is essential in the design, that is, the capability to operate at different production levels with acceptable efficiency levels.

5.1.2 Operability

Operations staff can provide information based on their past experience in running the plant. Using this information, the designer can design plants that are easy to commission, operate, and shutdown. Operators can help check these features while it is still in its early stages of design. In order to shut down plants safely, the operators' feedback can help identify special design features. Ergonomic considerations can play an important role in safe operations. An operational review of the three-dimensional model of the plant will take this into account. The costs and impact on the schedules of the resulting design changes can be quite low. Operational staff exposed to the design at an early stage become familiar with the plant long before the date of commissioning. This helps identify the gaps in their training and skills, which can be filled while the operators are still in their current jobs. Operator involvement can be a very motivating and satisfying experience. It will improve their pride and ownership of the plant.

5.1.3 Maintainability

The ability to restore a defective item of plant quickly is a measure of its maintainability. There are three issues to consider in the design in this context:

- It should be possible to locate the fault and identify the cause quickly;
- Access to the defective equipment or parts should be easy;
- Lifting gear, transport, and lay-down facilities must be available.

Modern photo copiers illustrate the use of improved diagnostic aids, including self-diagnosis. These machines tell us how to trace and rectify the fault when it occurs. Access to most parts is by operating simple clamps, levers, or hand whccls. Older generation machines did not boast of such features, and the improved maintainability will be evident to those who have used both varieties. The former Procurement Executive of the Ministry of Defence in the United Kingdom has produced an excellent video called 'Maintenance Matters' on defense equipment maintainability. In one example in this video, they compare two designs of fighter aircraft. There is a black box for recording the relevant flight information, in both designs. A technician removes the unit after each flight to download the data. The black box is in a compartment accessible from the outside, as illustrated in Figures 5.1 and 5.2.

In one design, the cover of the compartment is secured with about seventy fasteners.The fasteners have different types of heads. These include cross-head and high-torque screw heads as well as more conventional types, and with different sizes. As a result, the technician needs seven different tools to open the cover. Then he has to lift it out bodily and place it on the tarmac, before pulling out the black box. In the other design, the black box compartment cover is hinged along the top edge. It is secured by three toggle-clamps along the bottom edge. The technician can open the cover easily and quickly by operating the clamps. In the open position, the cover doubles as a rain protection.

The difference in maintainability in the designs will be evident from these two figures. The second design enables rapid retrieval of the black box, and the time required to do the work is only a small fraction of that required ear-

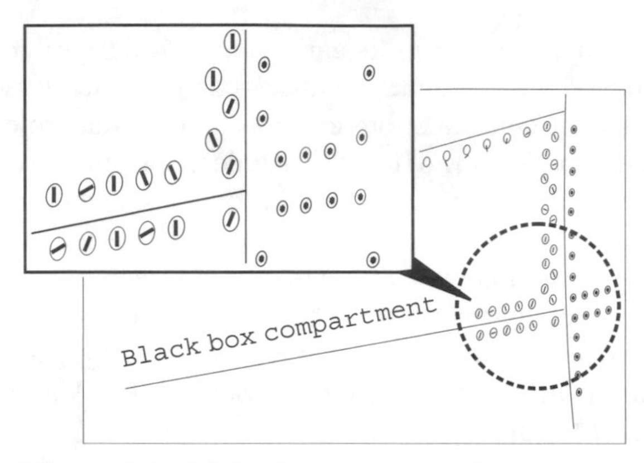

Figure 5.1 Multiple screw fastening system.

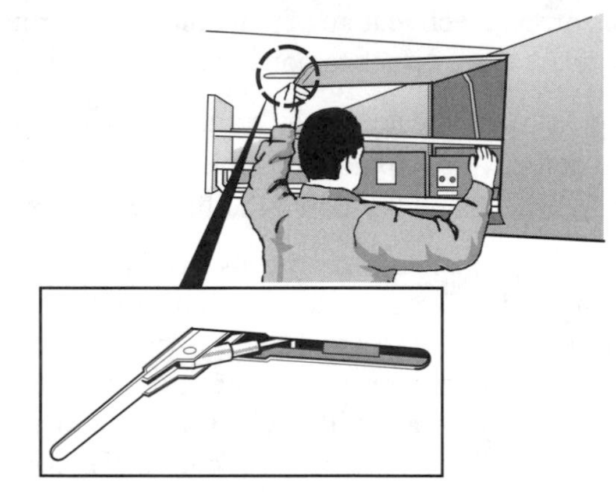

Figure 5.2 Hinged toggle-clamp system.

Figures 5.1 and 5.2 are reproduced from the video 'Maintenance Matters', courtesy of the Ministry of Defence, United Kingdom.

lier. Through the lifetime of the aircraft, the maintainers will enjoy the benefits of the additional thought and attention given to the maintainability aspects.

In the same video, they illustrate poor maintainability in another aircraft design. The example is about emergency batteries that need periodic servicing. In order to reach the batteries, the technician has to remove the ejection seat and the top of the instrument panel. Then he has to move the circuit breaker panel to one side, and remove a part of the rudder panel before reaching the batteries. Thereafter, the items have to be reinstalled in the reverse order. He does this work once in six weeks, so one can imagine his frustration and possible safety implications.

In an offshore oil platform, the author inspected a diesel-engine driven hydraulic pump. This provided motive power to a hydraulic turbine that was used to start up a gas turbine. The hydraulic pump and engine were on a compact skid, so tightly packed that it was very difficult to reach the instruments or critical engine parts. This remained a problem unit through its life.

In contrast to the previous figures, the photograph in Figure 5.3 shows a control panel in a modern offshore Floating Production, Storage, and Offloading unit (FPSO). Note the compact fold-away design of the computer keyboard, which allows easy access to the printed circuit boards.

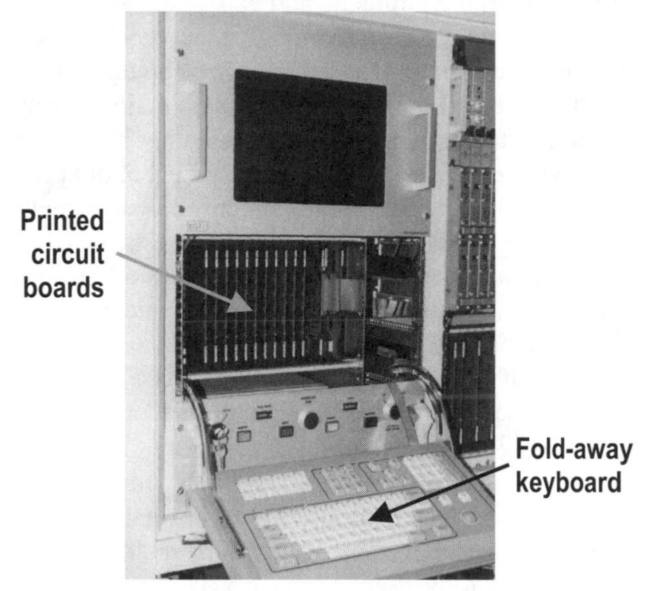

Figure 5.3 Control panel door.

The designer has to consider the range and volume of the anticipated maintenance work.We require adequate workshop facilities and lay-down areas with cranes and other lifting gear. The anticipated workload and availability of third-party facilities will help specify the requirement of machine tools. The main criterion in defining the size and location of the warehouse is the ease and speed of retrieval of spare parts. Contractors and vendors may own and operate the workshop and warehouse, if that meets economic and strategic criteria.

We can identify maintainability issues by reviewing the three-dimensional model of the plant. Maintainers are the best people to do this work, and they can suggest solutions as well. Software packages are available to simulate maintenance actions of male and female human models, if the three-dimensional model of the plant is available in electronic format. Using such packages, one can easily identify access and handling problems. This type of study will help reduce unnecessary downtime and maintenance cost over the life time of the plant. By solving the problems before commencing fabrication work, and avoiding needless change requests, we can save money and time. At the same time we can minimize the risks associated with their implementation.

Further discussion on this topic may be found in the book *Systems Maintainability*[1].

5.1.4 Reliability

We want reliable industrial equipment, and expect the vendor to build it into the design. As users, we do not generally give the vendor feedback on how well their equipment performs. Often there is no contact with a vendor and we make the first phone-call only when planning a major overhaul or after a catastrophic failure of the equipment. Vendors do not have access to operational history, but we expect them to know everything about the reliability of their equipment. Not having a crystal-ball they have to make intelligent guesses based on the demand for spare parts and requests for service- engineer support. The limited exposure during major overhauls or serious breakdowns is not enough to judge operational performance adequately. Without proper failure histories, it is difficult for equipment vendors to improve their products. Much of the fault lies with the user, but there is a lot more that vendors can do to gather failure data. Some vendors do manage to overcome these hurdles—but these cases are few and far between.

Another problem is that buyers of capital goods often do not specify reliability parameters in their requisitions. There are many reasons why this occurs. First, the measurement of reliability performance has to stand up to contractual and legal scrutiny. Second, buyers have preferred suppliers, for sound business reasons. These reasons include the standardization of spare parts, and satisfaction with previous support and service. Competitive prices or quality considerations do not govern whom we buy from any more, since the overall economics depend on such preferences. A vendor who has made great strides in improving the reliability of the equipment may still lose out to the established vendor. Hence reliability performance is an important selling point the first time we purchase an item, but thereafter other criteria become significant. Third, the actual buyer is often the design and construction contractor, not the ultimate customer who owns the plant. If the owner does not specify a detailed list of preferred vendors, the contractor will choose the vendor based on their own experience with different vendors. Once the customer and the vendor have to deal through a contractor, the importance of the views of the customer diminishes.

Contrast this situation with that of sellers of consumer goods and services. A manufacturer of a consumer durable such as a washing machine or an automobile sells the product directly to an end user, as do service providers such as airline companies. Even though there may be agents and intermediaries who handle the actual transaction, the deal is clearly between the manufacturer and the final customer. The marketing effort focuses on the end user. The two parties at the ends of the chain settle warranty or liability claims between themselves. Reliability now becomes important, because the customer wants it and can influence the supplier. If the customers are unhappy with the product or service, they can take their business elsewhere. Thus, in the case of consumer goods, the manufacturer makes every effort to keep the customer happy by providing reliable goods and services. When there is a direct link between the manufacturer and the ultimate consumer, customer preferences on product or service reliability assert their importance.

We noted earlier that some vendors find a way to collect failure history data in spite of the customer's unwillingness to oblige. For example, some vendors provide service centers for carrying out repairs. As a result, they have access to operating history; therefore, failure data becomes available and they are in a strong position to make reliability improvements. In most cases, these vendors are dealing in consumer goods and services, but there are a few cases of vendors of industrial equipment providing similar services.

A major manufacturer of printers has remained at the top of the market for a long time. Its products have a very good reputation. One of its customer service strategies is to provide convenient repair facilities for its units. One phone call gets you an agent to log your complaint. It offers a repaired unit to replace your machine or to repair it, if that is your preference. Then it transfers your call to a courier service that arranges to collect and deliver the units. There is no fuss, delay, or bureaucracy. The company retains customer loyalty, and should get excellent failure data from its service departments.

Industrial equipment buyers can use simple measures of reliability, for example, by specifying minimum run lengths between overhauls. The API 682 Standard—shaft sealing for centrifugal and rotary pumps—has taken a lead in this context. It states a design requirement of three years of uninterrupted service, while complying with emission requirements. This means that we can build warranties into the contract, with penalties for poor reliability performance. Once the general population of buyers starts specifying such requirements in their purchase orders, the suppliers will find a way to gather failure data.

A plant consists of many systems, sub-systems, and equipment items. From a reliability point of view, these may be in series, parallel, or some combination. In a series system, illustrated in Figure 5.4, failure of any one component will result in a system failure. For the system to work, all three components A, B, and C must work. In Boolean notation, we represent this by using AND gates to link the components.

Let us use the example of an automobile to represent a complete plant. In order to function properly, its engine, transmission, steering, suspension, and safety systems must all be in good shape. We show these systems with the blocks in series, similar to that in Figure 5.4. If we make a simplifying assumption that each of the systems' failures can be represented by an expo-

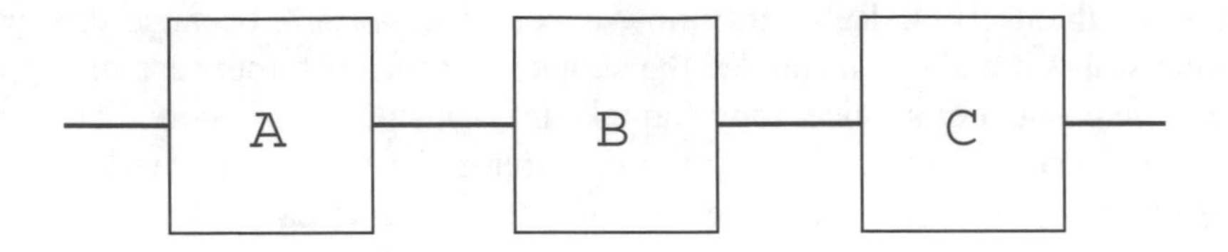

Figure 5.4 Reliability Block Diagram of a series system.

nential distribution, the overall plant reliability is the product of the individual systems' reliability.

Note that as the number of components in series rises, the system reliability falls. Figure 5.5 illustrates a system consisting of 20 components. For simplicity, we assume that each component has the same high level of reliability, ranging from 0.999 to 0.98. The corresponding system reliability is 0.98 in the best case and 0.667 in the worst case.

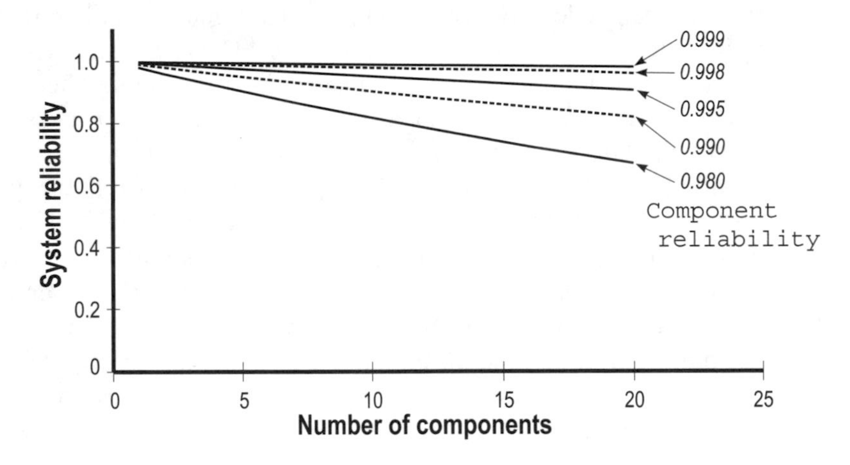

Figure 5.5 Effect of component reliability on system reliability.

This is one reason why complex systems are sometimes unreliable. Even when the component parts are very reliable, the overall system reliability can become quite low. This is an important lesson for designers of protective systems, which they use, for example, to safeguard critical equipment.

However, some designers make these systems very complex. This can be non-productive and, in extreme cases, positively dangerous. When there are many series elements (in terms of the reliability block diagram), there is a steep fall in the system reliability. We cannot ignore the so-called KISS principle (keep it simple, stupid!).

Figure 5.6 shows a reliability block diagram with parallel elements. In this case, we need only one of the components to work for the system to be effective. As long as A or B or C works, the system will work. Examples of such an arrangement are fire detection systems with voting logic, and standby equipment in a one out of two (1oo2) or two out of three (2oo3) or similar configuration. In Boolean notation, we represent this arrangement as elements connected by OR gates.

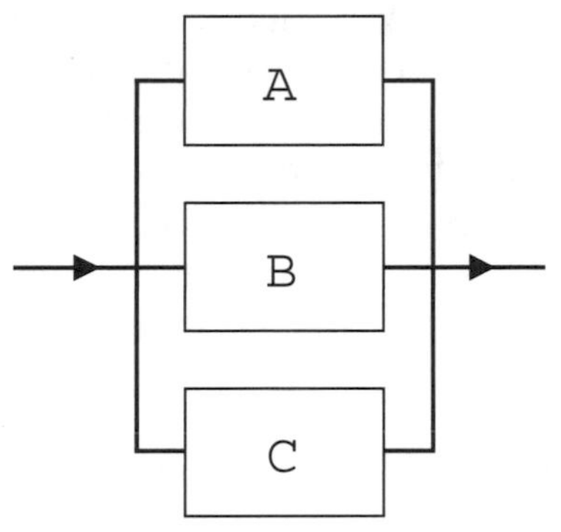

Figure 5.6 Reliability Block Diagram of parallel elements.

The system reliability increases very rapidly with the level of redundancy. With a high level of redundancy, we can tolerate very low component reliability levels. Figure 5.7 illustrates this observation for components whose failures follow the exponential distribution.

The reliability block diagram of an industrial plant can have a number of series-parallel combinations. The configuration reliability and capacity rating of each of the blocks representing the individual systems will determine how effective the whole plant will be in meeting its functional objectives. Some systems will have a bigger impact in terms of loss of function and are, therefore, more critical than the others.

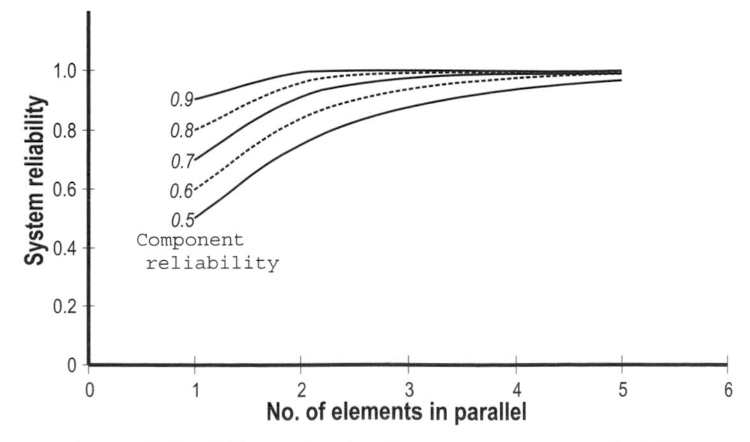

Figure 5.7 Effect of redundancy on system reliability

5.2 RISKS DURING CONSTRUCTION

The construction phase of a project is a period of high activity, often carried out under severe time pressure. We require a large workforce to fabricate and erect the plant. An added complication occurs when workers and specialists come from different countries, perhaps speaking different languages. The fabrication and erection work require heavy duty machines such as cranes, concrete mixing plants, bulldozers, and trucks. Weather conditions can be variable and severe. During construction, the plant inspection may require radiography, and hydrostatic and electrical high-voltage testing, with the associated safety concerns. In addition to these technical hazards, there may some other hazards that can be equally demanding, for example, outbreaks of industrial unrest, illness, or food-poisoning, or interruptions in the cash flow. The construction manager faces some combination of these risks during this phase.

A discussion about the risks of industrial action, public health, and cash flow is outside the scope of this book. The remaining risks in managing construction projects are similar to those encountered in process plant shutdowns, and are dealt with in Chapter 6.

5.3 THE PRE-COMMISSIONING AND COMMISSIONING PHASES

Prior to these phases of the project, the equipment is free of process materials. In a chemical process plant, the vessels, pipelines, and other equipment will be full of air or other non-corrosive fluids. A mechanical process plant will have coatings or other physical protection. There will be no electrical power supply to the plant, except that supplied for construction work. Thus the process itself is not a source of hazards. During the pre-commissioning of the plant, we prepare the equipment for service by internal cleaning, removal of preservatives, filling of lubricants, catalysts, and other chemicals. We also remove any mechanical locks, for example, those that vendors use to prevent movement of rotating elements. Some process plants need pre-heating or pre-cooling. We may use air or steam-blowing to clean equipment and pipelines internally. Some of these activities are themselves hazardous, but we can make suitable provisions to carry out the work safely.

We introduce process fluids during the commissioning phase. These may interact with the internal environment with potentially hazardous consequences. Where relevant, we must provide an inert environment so that we minimize such hazards.We must follow the vendor's start-up instructions closely in the case of complex mechanical equipment such as precision machine tools, heavy duty presses, large compressors, or turbines. It is advantageous to have the vendors' commissioning engineers present when we start such equipment for the first time. Their skills and experience will come in handy if we encounter unusual start up situations.

5.4 PLANNING OF MAINTENANCE WORK

Planning is the thought process to visualize the execution of the work. It takes place in the mind of the planner, who may use charts or other aids including computers, to do this work effectively. It is important to do this process as early in the design as is practicable. By doing so, the planner can ensure that the commissioning and startup activities are smooth, and that there is a bump-less transfer to the operating staff. Good maintenance planning, including the planning of spare parts, will go a long way in minimizing the risks in the operating phase.

Various planning tools are available, and we will discuss some of them in Chapter 10. Of these, Reliability-Centered Maintenance (RCM) is particularly elegant. One of the spin-offs in doing an RCM analysis is that it will help identify the failure modes where redesign is the only option to mitigate the potential loss. What this means is that the intrinsic design reliability is unacceptable and needs improvement. For example, in order to improve the system reliability, it may be necessary to install standby equipment in a system where the RCM study identifies unacceptable downtime. The benefit of doing maintenance planning at an early stage of the design is that we can do these changes on-screen or on the drawing board, well before fabricating and erecting the plant.As a result, safety, operating costs, and production volumes will all improve. If we do not use RCM or similar analysis tool, such redesign requests will only surface a few months after startup, when improvements are more difficult to implement.

5.5 THE OPERATIONAL PHASE

This is the phase in which the process plant will be for most of its life and exposure to potentially hazardous events is high. Even if the process itself is benign, the long period of exposure may mean that untoward incidents could take place. If the process is intrinsically hazardous, the probability of such an incident taking place becomes even higher. Maintenance work accounts for 35-45% of all the major injuries in the process industry. These happen during maintenance or as a result of wrongly executed maintenance[2].

5.5.1 Steady state operations

Some processes are intrinsically steady. The raw materials and other inputs arrive in a nice orderly stream, the production levels are constant, and the finished products leave the plant regularly. In such cases, the process fluctuations are minimal, and we call it a tram-line operation. With such a process, it is possible to predict the performance parameters fairly accurately. Good predictability will result in a high level of control. Such plants are likely to operate with fewer untoward incidents than others that have wide fluctuations in the process.

5.5.2 Competence and motivation

The knowledge, experience, and motivation of the operators and maintainers contributes significantly to the safety and efficiency of the production process. It is important to ensure that the staff employed to operate and maintain the plant are competent. We can and should measure knowledge and skill levels. Motivation is a complex issue, and we will discuss this further in Chapter 7. One has to work patiently and constantly to motivate staff.

People lose skills that they do not practice regularly, and forget theoretical aspects. This may affect their competence adversely. From time to time, we introduce new technology in the plant, either in the process itself or in the supporting infrastructure. Software upgrades take place continuously and at rapidly increasing frequency. High-performing companies carry out skills gap analysis and ensure that staff training fills the gaps.

We require proper documentation, drawings, written procedures, and work instructions to guide the staff. We have to keep them current, by periodic reviews and updates. If an electronic document management system is in use, it will ensure that the staff are able to see the latest version at all times. This will minimize the probability of different staff using different versions of the same drawing or procedure. When different versions of a procedure are in use, the probability of an untoward incident increases. Readers who have investigated accidents or major equipment failures will recall such situations. While carrying out a particular root cause analysis, the author found that there were three different versions of a steam-turbine start-up procedure in use. In this case, it was not the root cause of the failure, but the lack of control of important procedures was symptomatic of weak management systems.

5.6 MODIFICATIONS TO PLANT AND CHANGE CONTROL

We design and build new plants to well understood and accepted standards. Then we carry out various checks during the stages of construction to verify that the plant is safe to startup, operate, and shutdown. We carry out design reviews, hazard and operability studies (Hazops), and audits to verify the safety aspects of the design. It is, therefore, reasonable to assume that new plants are safe to operate.

Once the plant gets over its teething problems and gets into steady state operation, there is a drive to improve the profitability of the plant. Operators initiate change requests either to correct errors in the original design, or to de-bottleneck the plant. If we engineer these changes properly and think through the safety implications, there should be no problems in implementing them. In some organizations, proper change-control procedures may not be in place. Occasionally, there is a temptation to take short cuts to speed up the implementation. In other cases, operators may dream up temporary solutions to overcome pressing problems. The operators may not perceive their request to be a "change," and they may not apply the relevant procedures.

A simple rule to observe is that a like-for-like replacement does not warrant the use of a change control procedure. If we alter the materials of construction, physical location, or dimensions, move set points outside the design envelope, or rewrite software code lines, we must invoke change control procedures. Similarly, a change in process fluid composition is also a plant change.

Several industrial disasters have taken place due to use of inadequately engineered temporary solutions. One of the best known is the Flixborough[3,4] disaster, which happened on June 1, 1974. We will describe the incident itself briefly to recapture the main points.

In the first stage of the process used in the Flixborough plant, cyclohexane was oxidized at 8.8 kg/cm^2 and 155°C. The reaction took place in six stainless steel lined reactors connected in series. Each reactor was slightly lower than the previous one, by about 36 cm. Reactor No. 5 had leaked from a 1.8 m long crack some two months before the incident in question. In an attempt to continue production, the plant management decided to remove reactor No. 5, and connect reactors No. 4 and 6 directly. The difference in elevations meant that they had to offset the piping by about 36 cm, which was originally provided by Reactor No. 5. The nozzles on the reactors were 28" in diameter, but only 20" piping material was available. The connection between each pair of reactors required bellows, to allow for thermal movement. When reactor No. 5 was removed, the temporary piping (to connect reactors No. 4 and No. 6) had two bellows, one at each end, as sketched in Figure 5.8.

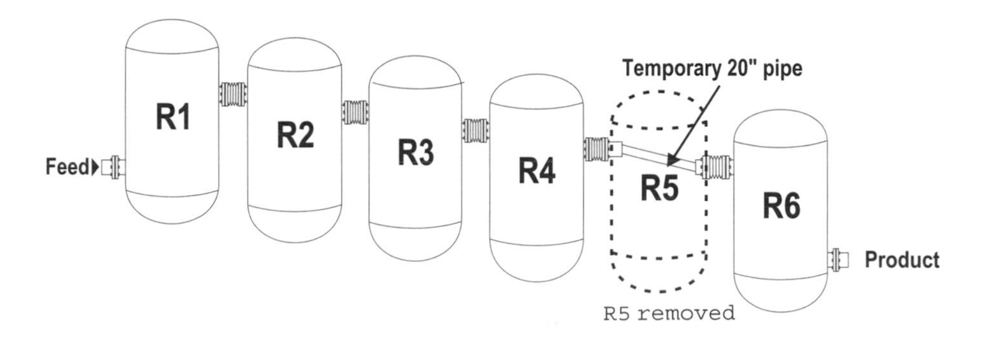

Figure 5.8 Temporary piping arrangement.

On June 1, 1974, the commissioning of the plant was in progress. There were some problems during the startup. At 4:53 p.m., an explosion took place, demolishing a large part of the plant. Twenty-eight people died and 36 people were injured. Fifty-three casualties were recorded outside the plant. An estimate of the size of the explosion was that it was equivalent to 15 to 45 tons of TNT. The design of these bellows allowed for axial movement but not for large angular movements. The 36 cm offset between the nozzles on Reactors 4 and 6 meant that the bellows would be subjected to excessive angular movement-, a fact not recognized in the design of the temporary piping.

The official Court of Inquiry concluded that the disaster resulted from a failure of the 20" temporary line. The responsibility for the temporary design rested with the works engineer, but at that time the position was vacant. The plant services engineer, whose background was in electrical engineering filled the post temporarily. The Court of Inquiry observed that the incumbent was not qualified to coordinate the work of the engineering department. The design of the temporary line did not conform to the relevant standard (BS 3351:1971) and the design guide of the bellows manufacturer.

The Court identified several other management failures. These included, for example, storage of 51 times the licensed capacity of flammable materials. There was no change control procedure in place. The reporting relationship of the safety and training manager was not clear. The Court of Inquiry held that there were failures in management resulting in the lack of a safety culture in the plant.

The lessons from Flixborough are clear; change control is important. In recent times, there is strong growth in the use of software in controlling the process. Software changes are considerably more difficult to validate and control. Use of object oriented software can simplify the validation process and is therefore worth considering up front.

High performance companies conduct periodic external audits of their maintenance and engineering systems. These will help identify any lacuna in change control procedures, and such audits are therefore recommended.

5.7 MAINTENANCE COSTS

The ratio of maintenance cost to total operating expenditure (Opex) can be quite high, ranging from 10 to 40%. As a result, this item of expenditure attracts a great deal of attention. People do not always recognize the contribution of maintenance in improving short term and long term profitability, but are invariably quite aware of its costs. The cost conscious plant manager needs a proper justification for the large sums used up in maintenance. The results of maintenance cost reductions take time to filter through, while the cost savings will be effective immediately.

A proper risk evaluation will identify whether doing maintenance has potential to improve safety or reduce production loss. The justification is by a cost-benefit evaluation. There are many reasons why maintenance managers do not produce such justifications. The unavailability of data, unfamiliarity with the methodology, and lack of time and resources are the most important reasons.

5.7.1 Failure rates and their impact

Let us now examine the impact of failure rates on Opex and revenue. We discussed some of the theoretical aspects of failure in Chapter 4, and noted that failures can be incipient, degraded, or critical. Since incipient failures exhibit symptoms of impending damage, we can plan and execute maintenance work so as to minimize loss of production. In this case, the adverse impact on profits is minimal. Degraded failures can result in reduced safety protection or a slow-down in production. It is possible to recover some or all of production losses by boosting production on completion of the repair. Usually, we can tackle degraded failures some time after detection, so this provides time to plan the work. Critical failures cause an immediate loss of function. Breakdowns and trips of critical items need immediate attention, as these failures may cause loss of integrity or production capacity. There is a penalty in terms of potential lost revenue or integrity. Failures result in direct maintenance costs, as well as loss of income during the period the equipment is unavailable. With high failure rates, the penalties become larger. The reduction of failure rates to the technical limit, namely the intrinsic or design levels, is therefore the best way to minimize these penalties.

5.7.2 Maintenance cost drivers—normal operations

Process plants need maintenance during normal operation, with associated costs. Maintenance costs are those relating to inputs such as materials, labor, energy, and supervision. We enter these costs into accounting systems, and often these are the only metrics available for control. If we delay maintenance work unduly, excessive damage may occur. For example, condition monitoring trends may indicate an incipient bearing failure. If we delay the repair to accommodate production pressures, it may result in the destruction of the bearing. In the worst case, it could even seize onto the shaft. This results in an increase in material and labor costs and extends the duration. Clearly the real cost driver is the delay, but this will not be evident from an examination of the cost records.

Similarly, poor operating practices can lead to avoidable failures. Some machines, such as steam or gas turbines, need a controlled rate of rise in speed. In the case of some pump seal designs, we need to balance the pressure in the seal chambers. In other cases, gas pockets in stuffing boxes require venting. The vendor manuals will state all these steps clearly, but people do not always read or observe them.

There are a number of reasons for poor operating practices. These include lack of ownership, time pressures, lack of training or motivation, and previous success in taking short cuts. We cannot find evidence of poor operating practices easily, especially when the events take place in the silent hours. On the other hand, if operators have ownership and pride in their work, they will take care to start-up and shut-down equipment in accordance with the vendor's instructions. They will also report deviations and errors, whether it is their own or that of others.

The quality of previously completed maintenance work affects the failure rate, as well as the ease with which we can carry out subsequent repairs. The skill, pride in work, data, tools, facilities, and the time pressures under which the technician has to work are important contributing factors. A technician who assembles a pair of flanges without lubricating the bolts creates a problem for the person who has to work on the same flanges later. *Maintenance managers who concentrate on cost or productivity alone, may involuntarily encourage poor work practices. Good quality work may cost more initially, but will pay for itself over the life cycle.*

There is an inherent failure rate associated with every piece of equipment. This relates to the design and construction quality, and how close the operating conditions are to the design envelope. Poor operating and maintenance practices make the actual failure rate worse than that built-in by virtue of the design quality. The difference in intrinsic and actual failure rates can be quite large, sometimes as much as ten times the ideal level. This gap offers the greatest potential for improving maintenance performance, and the first step is to measure and monitor failure rates.

In order to reduce the consequences of failures, we can adopt various maintenance strategies. Time or condition-based maintenance depends on the ability to carry out maintenance some time before the functional failure occurs. There is a penalty incurred, as in some cases parts will be replaced prematurely and their residual life lost. If the maintenance intervals are too short, we will incur additional costs or penalties. The issue is how well we are able to predict the timing of functional failure and that failures that do take place are analyzed, and causes established.

We incur maintenance costs when we execute planned work or when trips and breakdowns take place. The inherent failure rates, the quality of operations and maintenance, and the ability to predict functional failures determine the activity level. The efficiency with which we carry out this activity determines the cost. One factor affecting the efficiency is the productivity of the workforce and often this is the only one addressed. Improving work quality offers the greatest rewards, so this should always take precedence over efforts to improve the speed or productivity. Quite often, the actions point in the opposite direction, namely to reduce costs without an effort to monitor quality. *The correct solution is to eliminate or reduce the work itself before attempting to do it more efficiently.*

Nevertheless, productivity is an important issue to address. There are a number of reasons for low productivity, mostly caused by delays, rather than slow speed of work. Delays may be due to:

- single-skilled technicians
- policies and procedures
- low morale and motivation
- non-availability of parts, drawings, procedures, or instructions on time

Craft flexibility requires technicians to have one primary skill and one or more secondary skills. Waiting times can be reduced and productivity as well as job satisfaction can be improved significantly.

Management does not always realize that they may have policies or infrastructure in place that lowers productivity. Sometimes, permit-to-work procedures that are in place cause delays without adding to safety at work. Or the timing of breaks during the day may reduce work periods excessively. Sometimes reward systems do not support the drive to improve operational reliability. For example, elimination of trips and increase in run-lengths which should be rewarded may be ignored, while reduction of backlog may be recognized.

5.7.3 Maintenance cost drivers – shutdowns (turnarounds)

In Chapter 6, we will examine the way in which we determine shutdown intervals. We execute a very large proportion of planned maintenance work during shutdowns. These shutdowns also contribute to a significant proportion of the downtime. Reducing the frequency and duration of the shutdowns is an effective way to reduce maintenance costs over the life cycle of the plant.

However, longer intervals between shutdowns can result in more in-service failures, resulting in increased downtime. Hence a balance has to be struck and the optimum interval determined for each plant.

Maintenance work that cannot be carried out during normal operations, for reasons of safety, feasibility or economics, is done during periodic shutdowns of the plant. These shutdowns can be very expensive, often as much as two to five times the annual normal maintenance costs. In order to get an appreciation of the impact of shutdown intervals, consider the following example computation, based on an assumed set of costs. All figures are in millions of U.S. dollars.

a. Annual normal maintenance costs 15
b. Cost of shutdown carried out every two years 25
c. Cost of shutdown carried out every three years 30*
d. Cost of shutdown carried out every four years 38*

*Note: Increase in costs are due to larger scope of work, and includes costs of additional short intermediate shutdowns to cater for more equipment breakdowns during the larger intervals. In practice, work scope and costs do not increase in direct proportion to the extension in intervals.

In each of these cases, the annualized maintenance cost will be,

i. For 2-yearly intervals., annualized cost = 15 + 25/2 = 27.5
ii. For 3-yearly intervals., annualized cost = 15 + 30/3 = 25
iii. For 4-yearly intervals., annualized cost = 15 + 38/4 = 24.5

Shutdowns keep the plant idle for long durations, often accounting for 2-5% of annualized planned unavailability. The resulting lost production value can be very high, so we must make every effort to reduce this downtime to the extent possible. Can we extend intervals indefinitely? This is not possible for the following reasons.

1. The inspection interval of certain equipment is specified by national or state law.
2. As intervals increase, more breakdowns can occur. In chapter 10, we discuss methods to improve operational reliability, and thus reduce number of trips and breakdowns. However, there is a physical limit to these improvements. At some point, the downtime and cost of the additional breakdowns will be more than the gain due to the increased intervals. Each plant will thus have an optimum shutdown interval. We have to actively seek out this optimum and not accept the status quo.

In a similar manner, reducing shutdown durations increases uptime and often, but not always, reduces shutdown costs. Arbitrary reductions in durations are counter-productive. Reducing workloads, using better technology and tools, and improving staff motivation are the best ways to decrease durations. As examples of activities that can help reduce duration, we could,

a. Reduce shutdown work scope by doing as much maintenance as possible during normal operation, as long as it can be done safely and economically.
b. Do on-stream inspections to gather as much knowledge about the state of the plant. This will reduce surprises during the shutdown.
c. Use any opportunity that presents itself, e.g., an extended trip, to carry out work that will reduce the plant reliability and/or eliminate shutdown work. This requires a planning system that anticipates and prepares for such opportunities.

The best time to collect data to help improve the frequency and duration of future shutdowns is while a shutdown is in progress. This is when we get to see all parts of the plant which are inaccessible during normal operations. For example, we can record the severity of fouling in e.g., furnace tubes and relate it to operating conditions in the previous interval. With this information and the knowledge of future operating conditions, we can predict when the fouling of furnace tubes will become unacceptable. Such data gathering is preferable to using arbitrarily determined timings.

5.7.4 Breakdowns and Trips

Unreliable equipment and systems result in breakdowns and trips. They reduce availability and are expensive. Such events can rapidly escalate out of control. In starting a reliability improvement program, investigation of plant (or critical system or equipment) trips should be high on the list of actions.

The complexity of the protective systems can be a major source of spurious trips, so if the design is poor, such trips will plague the plant all the time. In chapter 8, we describe the Milford-Haven Refinery explosion. The investigation by the regulator, the Health and Safety Executive, resulted in a drive to reduce instrumentation complexity across the industry. This was aimed at minimizing spurious trips.

5.8 END OF LIFE ACTIVITIES

All plants have a given design life. However, by partial replacement of parts of the plant that have become inefficient or technologically obsolete, we can extend the life of the plant as a whole, sometimes indefinitely. There are exceptions, as in the case of nuclear power reactors, mines, or hydrocarbon reservoirs. In all these cases, there is a definite end of life, even though it might be considerably later than that expected originally.

We have to close down these facilities safely, and in an environmentally acceptable manner. Surplus materials need removal, and we have to restore the site to its original state. The risks involved in this phase are similar to those in the construction and operating phases. Environmental clean-ups pose additional problems, and pressure groups are likely to try to influence the outcome. The additional risks relate to perceptions, or what the public believes exists. It is not enough in these circumstances to produce quantified risk assessment study reports, as these do not address the problem of perceptions. Openness, transparency, and public consultation are often necessary. In Chapter 7, we will discuss the apparent dichotomy between perceptions and reality, and why decision-making does not always seem logical.

5.9 CHAPTER SUMMARY

In this chapter, we examined the risks associated with the various phases in the life of a process plant. The knowledge should help us in improving the way in which we address these risks.

The quality of design affects the intrinsic reliability of the plant over its entire life. Management must focus its attention on getting the design right, and ensure that consultations with key players take place. We highlighted the importance of keeping the design as simple as possible by examining the impact of complexity on system reliability. The construction and commissioning phases are periods of high activity and high exposure. In order to ensure that we anticipate and address these risks, careful planning and preparation are necessary.

In the operational phase, changes in design pose the greatest risks. The changes may be to correct design deficiencies, to increase plant capacity, reliability or maintainability, or to adapt the facilities to suit changing market requirements. We have to minimize the risk of safety, environmental, and production loss incidents, and need to manage change properly. Maintenance costs are under attack, and rightly so, since they form a significant proportion of the operating costs. If we are to reduce the risk of the baby being thrown out with the bath-water, we need an understanding of maintenance cost-drivers.

Finally, we reviewed process plant end-of-life activities.The associated risks are similar to those in the construction phase. Additionally, they carry a significant risk in the area of public relations.

References

1) Knezevic, J. 1997. *Systems Maintainability.* Chapman & Hall. ISBN: 0412802708.
2) Hale, A.R. et al. 1998. "Evaluating Safety in the Management of Maintenance Activities in the Chemical Process Industry. Safety Science vol. 28, no. 1: 21-44.
3) Kletz, T.A. 1998. *What Went Wrong?* Gulf Professional Publishing. ISBN: 0884159205.
4) Kirkwood, A. 1997. "Flixborough Revisited." The Safety and Health Practitioner. August: 30-33.

CHAPTER 6

Process Plant Shutdowns

A design life of twenty or thirty years is normal for a new process plant. In an ideal situation, the facility would operate continuously for twenty years or more from the time of commissioning, without any interruptions. The reality is quite different, with many factors contributing to a significant reduction in operating run times. Some maintenance work requires a partial or total shutdown of the process plant. In the context of process plant shutdowns, readers in North America may be more familiar with the term *plant turnaround.*

We do not intend to cover all aspects of planning, executing and managing shutdowns in this chapter. We are limiting our discussion to the management of risks relating to shutdowns.

6.1 FACTORS AFFECTING OPERATING RUN LENGTHS

These factors fall into two main categories, loss of integrity and loss of efficiency. If the plant goes out of control and it is not possible to correct it in time or shut it down safely, this compromises its integrity. Mechanical failures that result in loss of containment, or structural failures that result in equipment or building damage also compromise the integrity. The outcome of such incidents would be personal injury, environmental or equipment damage. Loss of integrity causes long term damage to the company's profitability, public image, and employee morale. If the incident is very serious, it may result in the loss of the business itself, as in the case of Flixborough or Bhopal. Loss of efficiency affects the flow rate, yield and/or quality of products. This will reduce the profitability of the plant. We discussed the need for maintenance action earlier, in Chapter 4 (refer to Figure 4.4).

6.1.1 Loss of integrity

Due to a deterioration in the strength of the materials, or the failure to operate as designed, we can lose mechanical integrity. This can be due to fouling, corrosion, erosion, or fatigue. Damaged components require repair or replacement; for example, corroded boiler tubes have to be replaced. Alternatively, the applied load may be above the designed level, subjecting the equipment to stresses that are beyond its capability. The load increase is normally due to process deviations, or external environmental factors that are outside the design envelope. When this happens some redesign is required to upgrade components to function under the new load conditions.

We use various protective devices to prevent the escalation of events. These may be pneumatic, mechanical, electrical, or hydraulic devices, such as trip relays, relief valves, or overspeed trip devices. There may be complex instrument protective systems to safeguard furnaces or large gas compressors. If any of these devices does not function when called upon to do so, we may lose the integrity of the protected equipment. Periodic testing reveals the state of the protective device, whether it is working or not; defective devices need replacement or repair.

6.1.2 Loss of efficiency

The performance of the process plant will deteriorate with time as a result of wear or fouling of the equipment. Catalysts or molecular sieves that may be in use will lose some of their activity over time. There will be a loss of product quality, a reduction in the yield or throughput, or an increase in energy or other resource inputs as a result of such deterioration. By measuring the trend in the loss or efficiency, we can estimate the point at which this will become unacceptable. At this point, we have to address the factors contributing to the loss of efficiency.

Fouling is one of the main contributors to loss of efficiency. It is usually not economically viable or even possible to eliminate fouling, which can be from one or more of the following sources:

- Contaminants in the process fluids themselves;
- Deterioration of the catalysts or molecular sieve beds causing downstream fouling;
- Scale, rust, or other products of corrosion;
- Fouling agents in the cooling media;
- Products of the chemical process;
- External factors such as dust or saline environment.

Some kinds of fouling can be reduced and run lengths increased by, e.g., cleaning of upstream storage tanks. In specific situations, drag-reducing chemicals can be added to reduce the effects of fouling on flow rates. These steps can be costly and/or cause other problems downstream, so a proper risk evaluation should be conducted up front.

If we are able and willing to spend the money, we can design out some of the causes of failure due to corrosion, erosion, or fatigue. In many cases, it is uneconomic to do so, and most designs allow for some parts to fail well before the design life of the plant as a whole. There will always be opportunity windows, for example, to replace catalyst beds that have a finite life. Some equipment or component parts may be considerably less durable than the plant as a whole, because this happened to be the most economical solution.

It may be possible to predict revealed or evident failures by monitoring the equipment's condition. If the failures are unrevealed or hidden, it is not possible to find the time of failure, as discussed earlier, in section 3.7. Designers use instrument protective systems to protect process plants. These detect unsafe conditions or unacceptable process excursions, and initiate a controlled shutdown. Unfortunately, many of the failure modes in these systems are unrevealed. Such unrevealed failures will compromise the integrity of the plant. In order to detect failure, we need to carry out tests, which may require a plant shutdown.

If it is possible to predict the failure, it is a good practice to find an economic window of opportunity to carry out maintenance work. This is the logic of carrying out time-based or condition-based maintenance.

Instrument protective systems sometimes send trip signals even when there is nothing wrong with the process, resulting in unwarranted spurious trips. In Chapter 5, we saw how the unreliability of these protective systems relates to the complexity of the protective instrumentation (refer to Figure 5.5). The more protective instruments there are, the greater the chances of spurious trips. There is a drive to reduce the number of spurious trips by replacing some trip actions with alarms. There is also a move to reduce unnecessary alarms (refer to section 5.7.4).

6.1.3 Incorporation of plant changes

We need to incorporate various design changes to meet changes in market demands, improve operability and maintainability, or eliminate the root causes of premature failures. These change projects normally require a shutdown of the plant for carrying out the modifications.

In Chapter 5 we discussed the Flixborough disaster and the importance of change control. In this context, any deviation from the design intent is a change, including, for example, changes to pump impeller sizes, software, or relocation of valves.

6.2 RISKS RELATED TO PLANNED SHUTDOWNS

Planned Shutdowns give us an opportunity to restore the integrity, productivity and product quality of a process plant. The concentration of resources, intensity of supervision and involvement of senior management can help improve the quality and productivity of the maintenance crew.

Shutdowns are periods of high activity. There are certain risks involved in executing the large volume of planned work. These risks relate to personal injury, environmental or equipment damage, and of not being able to complete the shutdown work in time or within budget. We have to guard against poor quality work and the associated problems.

6.3 PLANNING

We defined planning as the process of executing the work mentally. We visualize all the steps required for each item of work, the resources, materials, tools, procedures and equipment we need, as well as the hazards we may face. We then identify the steps to mitigate any adverse consequences, including delays to the work. On the basis of this knowledge, we decide how we will do the work; that is then the plan. The thinking process enables the planner to visualize problems and to seek solutions on screen or paper. By doing so, planners can eliminate or reduce surprises. They evaluate various scenarios, and make suitable provisions to cater for all credible ones, thus ensuring an acceptable outcome. The amount of money spent per day of shutdown is often comparable to that spent in large construction projects. The difference in planning effort between large projects and shutdowns is quite striking. Projects often use planning resources that are an order of magnitude larger than that available to the shutdown manager. We have to address the question of whether we spend the right effort in planning shutdowns on a case by case basis.

A level of generic planning is possible even in the case of some types of unplanned shutdowns. In these cases, we can define the work-scope broadly, but the timing is unknown. The situation differs from plant to plant, and we have to identify the scope in each case. Some examples of potential causes of unplanned shutdowns are, for example, failures of furnace or boiler tubes, relief valves that do not re-seat after operation, and premature damage to catalyst or molecular sieve beds. We can define the scenario in each case and make a contingency plan, ready for use in a real situation.

The planning of shutdown work is a method to reduce the risk of an undesirable outcome, and improve the chances of achieving a successful one. We will discuss the focus areas for such planning in the following sections.

6.4 SAFETY AND ENVIRONMENTAL HAZARDS

There are a number of procedural steps that we can take to minimize health, safety, and environmental incidents. The aim is to reduce the occurrence of untoward events. In spite of these efforts, some events may still take place. If this happens, we try to reduce personal injury or environmental damage. A few of these events, if not adequately controlled, may escalate into major incidents such as fires or explosions. The plan must show how we can manage the situation, and limit the damage to an acceptable level.

In practical terms, we prepare a safety and environmental plan, to advise shutdown personnel. The normal shutdown work needs some preparation, and the plan should include the following items.

6.4.1 Traffic safety

Contractors usually provide most of the large work force required for shutdowns. The numbers involved could be in hundreds, and transporting these people from the gate can create traffic hazards when they have to walk to the shutdown site. It is quite important to evaluate these risks and plan accordingly. As an example, it may be necessary to separate vehicular and pedestrian traffic, and reduce the probability of accidents. For this purpose, the planner can designate some roads for use exclusively by pedestrians, and others for vehicular traffic, so that the two flow streams stay apart. Alternately, one may use buses for transporting people between the plant and the gate or canteen. The approach is similar to that used by airline companies, who use buses to transport people across the tarmac, even when the distances involved are small. This minimizes the probability of uncontrolled movement of the passengers resulting in security problems and accidents involving pedestrians.

6.4.2 Waste management

If we can eliminate or reduce the volume of waste economically, we should try to do this as far as possible. In order to minimize the risks in dealing with the rest of the waste, we suggest the following actions.

The first step is to make an inventory of waste materials and effluents that we may expect to handle in the shutdown. The next step is to estimate the volumes, and make arrangements for labeling, storing, handling, and disposing of solid, liquid and gaseous waste materials. Thereafter, the planner identifies procedures for handling these materials. The next step is to identifiy the person responsible for managing the waste materials. The plan must clearly identify fall-back positions that we can apply if the volume or toxicity of the waste products differs from that originally estimated. We have to make physical arrangements to manage the waste products and communicate the information to all the relevant parties.

6.4.3 Hazardous materials management

Certain materials such as asbestos and mercury require controlled handling.In chemical plants, one may encounter materials such as hydrogen sulfide, mercaptans, aromatic hydrocarbons, strong acids, or alkalis. The planner should identify such materials and prepare suitable procedures to minimize untoward incidents. The actions are similar to those in 6.4.2 above.

6.4.4 Fire and evacuation drills

Shutdown managers are invariably under extreme time pressure. It is difficult to justify fire and evacuation drills, which can be costly in terms of lost time and productivity. Induction and training programmes should prepare the workforce to react sensibly if a fire or toxic gas leak occurs. *However, the only way to confirm that the people understand the message is to conduct one or more drills.* The real question, therefore, is whether we can afford not to conduct such drills. We can minimize the loss of productive time by choosing the time we conduct the drill carefully, for example, immediately preceding a planned meal or tea break.

6.4.5 Tool box meetings

Prior to the commencement of work, the maintenance supervisor normally discusses with the assigned crew the main safety and technical issues relating to a piece of work. These meetings are of short duration, and held at the staging point, where we store tool boxes. *It is necessary to institutionalize these meetings, as they improve communications and reduce the probability of a safety or environmental incident.* They also help in improving the quality of the work, as the workers have a clearer idea of what they are about to do, and the purpose of their effort.

6.4.6 Emergency communication conventions

Every plant uses its own convention in operating sirens and alarms. Sirens may operate continuously or intermittently to indicate different types of unsafe situations. In some locations an intermittent siren indicates a gas leak, while a continuous siren may indicate this condition in another location. In some plants, colored flashing lights indicate the types of gas leaks. The responses to each of these unsafe situations can be different, so it is important to ensure that the work force understands their meaning clearly. The contract workers may not know the communication conventions in use in the plant in question, as they differ from one plant to the next.

It is necessary to explain escape and evacuation routes and identify muster points clearly. The direction of the prevailing winds will help decide where people should assemble in the event of an alarm. The shutdown crew should know about wind socks if any are in use. *An induction program designed to explain the plant-specific alarm communication system and the correct response is the best way to ensure a clear understanding.*

6.4.7 Training

Emergency procedures are useful tools in reducing the risk of injury or environmental damage. Adequate training of the relevant people is essential if these are to be of use. In an emergency, rapid evacuation of personnel from the affected area is important. They must leave the work-site in a calm manner. Evacuation routes, assembly points, and the operation of escape equipment must be clear to the people involved. There can never be too much of such training. All the commercial airlines, for example, explain the emergency procedures prior to every take-off. The shutdown crew must also know whom they should report to at the muster point so that a quick head count is possible. The designated leader at the muster point should be clearly identifiable. *How well the shutdown crew respond when fire and evacuation drills are conducted demonstrates the success of the training*.

6.4.8 Rescue planning

Rescue equipment should be available within easy reach of the shutdown personnel. These include, for example, lifting cradles, resuscitation equipment, breathing air bottles, rescue ropes, and lifting tackle. Designated rescue contacts need specific rescue training. Hazardous activities, for example, those needing the use of breathing apparatus, need additional precautions. The following check-list applies to hazardous work that does not require vessel entry.

- Provide two separate escape routes for the crew. If they need to climb up or down from the place of work, consider providing them with temporary staircase accesses;
- Have a crane and a lifting basket available on standby;
- Ensure that the local clinic is aware that a hazardous job is in progress;
- Prepare the work so that speedy execution is possible;
- Ensure that the work crew is aware of the hazards, escape routes, and procedures;
- Carry out a dry-run before commencing the work.

Work inside vessels is specially hazardous, and the crew must follow all the precautions stated in the permit-to-work document. Workers inside columns and vessels may suffer falls or be asphyxiated. The rescue facilities and training must cater to this eventuality.

The operators isolate the plant from the upstream and downstream facilities after shutting down the process. Then they isolate the electrical, steam, and other utility connections. Process and steam lines need positive isolation. Technicians insert spades (paddles or full face blinds) between pipe flanges to provide positive isolation, and this work can be hazardous. In a chemical plant that handles toxic or flammable fluids, these hazards are even higher. It is quite important to recognize and address the hazards relating to plant isolation and de-isolation. The planner must prepare a list of hazardous activities of the types discussed above; the relevant supervisor should discuss these risks with the work crew during the tool box talks. A rescue plan must be on the agenda of this discussion, and suitable preparations have to be in place to cover any eventualities.

6.4.9 Medical support

When the work is of a hazardous nature with the possibility of serious injury, it is a good practice to ensure that medical support staff are available on a standby basis. Emergency vehicles such as ambulances, fire trucks, and cranes may have to be available at short notice. We may never call on these services, but should nevertheless have them available on demand.

6.4.10 Reference booklet

Finally, a reference booklet should be issued, giving the names and telephone numbers of key personnel responsible for safety and environmental management.

6.5 WORK SCOPE AND ASSOCIATED RISKS

Work scope changes are the most common reasons for loss of control on the duration and cost of shutdowns. As discussed earlier, we carry out plant shutdowns for specific reasons. In order to control the volume of work to be included in the shutdown, we exclude work that we can do during normal operations. Good on-stream inspection techniques can help greatly in identifying essential maintenance work. Usually, we compile the work list for a shutdown eighteen to twenty-four months in advance. We prepare a detailed plan only after challenging, justifying, and accepting the work-list. We have to demonstrate that it improves the efficiency, reliability, or integrity, or relates to an agreed plant change. We have to eliminate items that do not meet at least one of these criteria.

6.5.1 Freezing of work scope

The next step in controlling the volume is to freeze work scopes as early as possible. It is a good practice to do this six to eighteen months prior to the shutdown. Obviously, this period would vary depending on the type of process plant under consideration. Once we freeze the scope, further changes need approval from the management team. Fluid work scopes are a sure recipe for disaster and they are likely to increase the duration and cost of shutdowns. In order to minimize these risks, management should keep a firm control on the shutdown work scope.

6.5.2 Work scope changes during the shutdown

Once they are under way, one can anticipate some changes in the planned work volume in most shutdowns. Such changes are likely to impact adversely on the duration and cost. We have to challenge all new work, whether it is contingent or emergent in nature.

Contingent work is that which we can anticipate, but needs additional data to confirm the earlier predictions. Wall thickness readings taken on a process pipeline during operation may indicate that a section needs repair or replacement. These readings need reconfirmation nearer the start or even after commencement of the shutdown. The new readings may either confirm the earlier decision, or indicate that we can postpone the work to the next window of opportunity.

Emergent work is that which we cannot reasonably anticipate. Usually, internal inspections reveal equipment damage that we cannot identify otherwise. As an example, the inspection of the internals of a column may reveal fatigue or weld cracks. If these are critical to the integrity of the column, we have to include the additional work to the plan.

The shutdown plan must provide for a volume of emergent work, even though details of such work are not available. The inspector can confirm or modify the actual work content on completion of the inspection. It follows that we should schedule the inspection of equipment that may yield emergent work early in the shutdown.

It is a good practice to identify those items of equipment, piping or other facilities that have the potential to raise emergent work. These items should be opened and inspected as early as possible in the shutdown. This gives the inspectors adequate time to define the scope of repair. We also need time to arrange resources, spares, materials, tools and logistics support.

In order to minimize the impact of work scope changes, the shutdown plan should have provisions for contingent and emergent work. Usually, we can define contingent work in advance. Emergent work is much more difficult to define, and the planner makes provisions based on knowledge of the process, and operating experience. Even though a provision is available, we must always challenge any additional work and, once approved, include it formally in the plan revision. There is a possibility that Parkinson's law (work expands to fill the time available) may apply, and provisions for emergent and contingent work used as a cover to hide inefficiency.

6.6 QUALITY

The saying, 'work worth doing is worth doing well,' is certainly applicable to shutdowns. We invest considerable effort in generating, justifying, and planning shutdown work. It is at least as important to ensure that we do the work to acceptable quality standards, by proper follow-up actions. The risks associated with poor quality work, namely delays, rework, and cost increases, can be minimized by paying adequate attention to the following.

6.6.1 Quality targets and performance indicators

The targets must be simple, direct, and easy to measure. We need just a few indicators, but these must be objective and measure final outputs. In a chemical process plant, any leak is potentially dangerous. They are usually easy to locate and difficult to dispute, so the number of leaks at startup can be one metric to use. A smooth start up is also easy to identify. Any defective workmanship observed during the shutdown, and recorded as a non-conformance is a suitable quality measure. We can apply similar performance measures in mechanical process plants as well. These must be appropriate to the kind of process and seen to be fair. We have to measure performance during the shutdown and startup, and communicate the results to the work force. The contract may include incentive payments to ensure that the contractor meets these targets.

6.6.2 Competence

As discussed earlier, the contractors employ large numbers of people to execute shutdowns. Many of these people may be new to the plant, and their skill levels unknown. A large proportion of quality problems is attributable to the lack of skills and competence. The effort involved in checking the competence can be high, and it is not a practical proposition for the shutdown manager to undertake this activity. However, it is worth asking the contract firms involved to do so, and we can include this requirement in the scope of work. Some additional costs will be incurred, but this can be easily justified by the expected improvement in work quality

6.6.3 Records and traceability—positive material identification

Some of the materials, fittings and spare parts may be in services that are very corrosive, erosive or have the potential to cause embrittlement or cracking. While the designer will take care to specify the materials to be used very carefully, it is absolutely essential to ensure conformity, and to be able to demonstrate that we have complied with the requirements.The manufacturer uses stricter standards for these items, and tests them more rigorously. They may require special heat treatment. We have to track these special materials, for which purpose we need proper records. These traceability records are important quality assurance documents, and we can also use them as a measure of the quality of work.

The shutdown manager must be able to demonstrate that the materials used during the shutdown in critical services are of the approved standard and that all the records relating to such work are properly documented.

6.7 ORGANIZATION

An important risk relating to shutdowns is in not being able to staff it with the right people, both on the Company's side and in the Contractors' organizations. It is critical to get this right, so preparations must start some 18-24 months before a major shutdown commences. The key positions in the Company's shutdown organization have to be named, and the incumbents released for the planning and preparation work. The shutdown manager should normally be working on the job on a full time basis for at least 18-24 months. Most of the other people will have two roles, a day-job doing their normal work and an additional job, preparing for the shutdown. The shutdown organization will have people from Operations, Maintenance, Inspection and when plant changes are involved, Engineering as well. Needless to say, management support is essential to ensure that their shutdown roles are not swamped by their routine work. Planning of operational activities is very important and must be fully integrated with the main shutdown plan. These include, e.g., shutting down, emptying the plant of process fluid and chemicals inventory, positive isolation of process streams and utilities. Other preparatory work such as tagging of items of work, preparation of work-permits, waste disposal procedures etc. are usually handled by the Operations planners.

The main contracts have to be awarded about 12-18 months before the start date, and the Contractors' key staff identified and approved. Good liaison between the shutdown manager and the Contractors will ensure that the latter rapidly designate and fill the remaining slots in their shutdown organization charts.

Internal transportation, canteen facilities, medical support etc. also need 3-4 months of advance planning and preparation. A person in the shutdown organization must be assigned the role of coordinating these activities.

6.8 EXECUTION

6.8.1 Safety aspects

In section 6.4, we noted the planning effort we need to make the shutdown safe. In practical terms, we have to manage hazardous work, and this is obviously more than just planning the work safety aspects.

We will be using cranes and lifting gear throughout the shutdown. Their spatial movements, overturning moments, supporting soil stability, visibility of load to the crane-operators, and communications between the rigger and the crane-operator are all important issues to control. Similarly, we will use welding machines and also carry out other hot work. The permit-to-work provides the procedural barrier to ensure that there are no flammable materials in the vicinity of the hot work.

There are hazards relating to vessel entries, dropped objects, scaffolding rearrangement, slippery floors, use of incorrect materials (see section 6.6.3), working at heights, congestion caused by machinery and people, housekeeping etc. Good supervision, attention to detail and a clear safety focus will all help keep these and similar hazards at bay.

Operations manage the process of shutting down and starting up the plant, isolating and de-isolating it safely and efficiently. They also keep a watchful eye on the (very) large maintenance crew, often from two or more contractors, to ensure that they work safely. They support maintenance by providing fire-fighting cover when needed, as well as in other safety roles. Good cooperation between operational and maintenance staff is essential for a safe and efficiently run shutdown. It is also essential for keeping the risks as low as possible.

6.8.2 Competence

The scheduler would have decided the crew size and composition based on the work content and duration. Shutdown work is intense and consumes large quantities of human and other resources. As a result, it is not uncommon to find a dilution in skill levels. The contractors must demonstrate that their workforce possess the required skills. This can be done by offering records of recent qualification tests conducted by independent agencies, or by arranging to conduct such tests before the shutdown, duly witnessed by the shutdown manager's representative.

6.8.3 Overlaps and interference

At any one time, several skill groups will be working on different tasks. Some of these will have a bearing on others in progress at that time. For example, inspection must follow cleaning work. Similarly, mechanical repair work follows inspection. Some expensive resources such as cranes may be in short supply, and we must schedule their movements carefully. When inspectors wish to carry out radiography, it is necessary to cordon off the affected work areas so as to protect other workers from exposure. All this needs good coordination. The additional complication is that different contractors may be providing the required services, so it is important that there is good communication between the contractors and company staff.

6.8.4 Productivity

Good planning and scheduling are an essential pre-requisite to attain high productivity. If the planner arranges the work periods and rest breaks properly, and the workers do not have to wait for instructions, permits, drawings, materials, cranes, or other resources, their productivity should be satisfactory. In a study conducted during a major shutdown, the author found that nearly 80% of the loss of productivity was due to poor planning, scheduling, and supervision. The premise that workers are inherently lazy is usually incorrect. However, if there is evidence of low productivity and attributable to the workers' attitudes, the manager should take disciplinary action.

The author has audited the execution of a shutdown where the gap between start-of-work, meal/tea breaks and close-of-work was such that the technicians were unable to attack the work for meaningful lengths of time. In this case, two changes were proposed,

1. Increase work day from 8 to 9 hours by adding an hour of over-time.
2. Arrange for Supervisors to start 30 minutes before the technicians arrive, so as to get all the work permits issued by the time the technicians come in to work.

With these changes and by adjusting the timing of breaks, 36% more working time became available. Additional costs related to the 12.5% increase in working hours and the (1.5 hour) longer day for supervisors, offset some of the labor cost saving due to the gain in available working time. However, the real gain was in reducing shutdown duration. In theory, this could be up to 9 or 10 days in a 30-day shutdown. In practice, we expect a 4-5 day reduction. The value of these improvements was in excess of US $ 6,000,000.

6.8.5 Closing-up equipment

Tanks, vessels, columns, furnaces, and other enclosed spaces where people have worked can create special hazards. An operator must sign a close-up checklist and supervise the final closure of such equipment. Each company must evaluate the hazards and make its own checklist. We offer the following points for consideration:

- Inspector confirms that repairs are complete and satisfactory;
- Operator confirms the removal of debris, surplus materials, tools and scaffolding;
- Operator confirms that internal parts are installed correctly;
- Operator confirms that all workers have exited the enclosed space;
- Operator supervises closure of each opening ensuring that the relevant flanges have at least 4 bolts that are hand-tight;
- Operator posts 'Not safe to enter' signs at each opening;
- Maintenance supervisor takes over, after Operator leaves, to ensure that all bolts are installed and fully tightened immediately thereafter.

A good rule of thumb to follow is to ensure that we start closing-up equipment about half-way through the shutdown. As far as possible, we should complete the closing-up of all equipment about two to three days before the scheduled start-up date. Pipe work and instrument re-installation should follow the equipment closely, so that the unit is mechanically complete at least one day before the start-up date.

6.8.6 Area clean-up

The shutdown crew must complete site cleaning and removal of surplus scaffolding and other engineering materials before the plant can be started-up. A pre-startup audit will establish that the plant is clean and ready.

6.9 SPECIALIZED EQUIPMENT OVERHAULS

The vendor's specialist engineers provide two distinct services. The first is their skill and knowledge that will help us carry out the overhaul work to the required standards. Their skill will also assist us during the start-up of the equipment. If during the overhaul, we encounter some unexpected damage, and we need additional spare parts, this can delay the start-up. If the vendor specialist is present, we can minimize such delays. They know who to contact and where the part may be available, so their presence serves as an insurance policy. This is the second service they provide.

6.10 COST CONTROL

In a large shutdown, the daily spend can be $1,000,000 or more. It is important to track the estimated, committed and actual cost on a daily basis. A dedicated cost engineer or cost engineering group is often required for this work. In smaller shutdowns, this may be done by the scheduler. The shutdown manager will check the variances and take corrective actions. Work progress and costs must be reconciled, and for this purpose the scheduler must keep track of the following.

- Work progress;
- Actual resources used versus the plan;
- Spares and materials consumed;
- Third party commitments.

Delays in completion of individual activities can affect the overall duration. The scheduler updates the plan daily or every shift, depending on the duration of the shutdown. These updates will show the impact of delays of individual activities on the overall plan. The shutdown manager must decide whether to divert resources from other less critical activities or provide additional resources to overcome such problems.

As in any other project, the scheduler must produce cumulative progress and cost curves at the same frequency as the plan updates. The shutdown manager can exercise control by tracking the costs in relation to the progress.

6.11 COMMUNICATION

During shutdowns, we use a large number of company and contract staff. The relevant supervisors require and provide information of use to one another and to the shutdown manager. A scheduled meeting every morning, shortly after the workforce commences work, is a good forum for information exchange. Towards the end of the day, the scheduler can meet each supervisor in turn, and update the plan with progress and cost information.

Technical issues will crop up, and these need prompt resolution. The shutdown manager will convene impromptu meetings to resolve such issues.

Senior management will also need progress updates. They may visit the site from time to time and request progress reports from the shutdown manager. Safety, cost, and time overruns are their principal concern.

The workforce needs feedback about safety incidents as soon as possible. Similarly, they need to know of any change in the situation that could affect their personal safety. Their supervisors will communicate most of this information in the tool-box talks.

6.12 CONTRACTORS

Some of the contractors provide resources and management skills, while others provide specialized services which are unique and not available within the company. Since we may employ many contractors in a shutdown, we have to take care of the interfaces between them as well as that with the company. Normally the contractors deal with their staff welfare and discipline issues. However, only the shutdown manager is in the position to keep control of the overall situation. If the contractors use specialized equipment or processes, they are the only people with the knowledge and experience required to manage the technical aspects of their work. The permit-to-work system helps maintain overall control.

6.13 SHUTDOWN REPORTS

These reports capture the history and the main learning points. The inspectors must recommend the timing and work scope for the next shutdown, based on their findings and the expected operational severity. The scheduler should advise the changes required in the plan, based on the data gathered and the learning points. The report should record the estimated and actual cost and resources, duration, work scope changes, and other relevant data.

The draft report should be available for review within 3 weeks of startup of the plant. It should be possible to produce the main report with the points listed above within 15-20 pages, and have all the supporting data in suitable appendices.

6.14 POST-SHUTDOWN REVIEW

The key players in the shutdown should meet three to four weeks after the start-up, by which time the first draft of the shutdown report would be available. We carry out this post-mortem to distill the learning points, which are useful for future shutdowns. The final shutdown report should be issued within a week or two of the post-shutdown draft report.

6.15 CHAPTER SUMMARY

Planned shutdown work, carried out at the appropriate timing, ensures that we obtain optimum performance of the plant. There are some risks associated with their execution. These arise as a result of changes in the work scope, variability in the quality of work due to competence and skills' deficiencies, inability to predict failures accurately, inadequate preparation, and safety incidents.

We examined how we can anticipate these problems and take steps to minimize the risks. Most of these steps follow the dictates of common sense and do not need special technology or tools. However, if we do not give adequate thought to these issues, we cannot expect optimum performance.

We can evaluate and make provisions for anticipated safety and environmental hazards. We examined issues relating to traffic safety, waste management, and the handling of hazardous materials.

Work scope changes pose a serious challenge in the control of the duration and cost of shutdowns. We can minimize this risk by freezing the work scope early and controlling changes to it during the shutdown.

The quality of work has to be of acceptable standard, otherwise there will be delay, rework, and cost increases. We need simple, easy-to-measure and meaningful quality targets. By testing the workforce, the shutdown manager can confirm their competence, thus satisfying one of the prerequisites for good quality work. Materials traceability records are important quality assurance documents. We discussed the importance of timely finalization of the organization and staaffing of the shutdown. All of the planning and scheduling effort is of no avail if we do not execute the work properly. The main issues of concern are safety, quality, management of interfaces, and possible extension of duration.

We discussed the role of the vendor specialist engineer in the overhaul of specialized equipment.

Normal project management tools and techniques are applicable to shutdowns as well. The scheduler tracks progress, commitments, materials used, and resources consumed. These enable the shutdown manager to control the duration and cost.

Managing shutdowns can be quite complex because of the large number of people involved, and the resulting interfaces. It is necessary to have good communication between all the relevant parties.

We examined the need for shutdown reports and post-shutdown reviews. These are the best instruments available to improve performance in future. The reports also form the basis of the work program for future shutdowns.

CHAPTER 7

Facets of Risk

Risk is a much misunderstood and sometimes misused word. This is not surprising; in its common usage in English, it can also mean chance or gamble. In insurance and financial circles, the word describes an asset, a person or financial instrument. The meaning of the word can therefore change with the context, and with the background of the people using the word.

In this chapter we will examine risk in its sense of loss of life, property or production capability. Whether we realize it or not, we make decisions based on our evaluation of the risks. It is therefore useful to examine and understand the facets of risk and we will cover the following:

- Our perception of risk affects the way we make decisions, so it will be useful to understand the relevant issues;
- In a quantitative sense, risk has two distinct elements: the probability or frequency and the consequence or severity of the event;
- The exposure or demand rate affects the required performance level. If one can reduce the demand rate, one can lower the acceptance standards without affecting the level of risk.

7.1 UNDERSTANDING RISK

The dictionary definition of risk is

1. *Risk: the possibility of incurring misfortune or loss...danger, gamble, peril, hazard.*
2. *To take a risk: to proceed...without regard to the possibility of danger....*

Collins Dictionary and Thesaurus, Harper Collins Publishers

Note that the stress is on the possibility of occurrence of events. In day-to-day use, the term *risk* is used in the sense of probability of the event.

Strictly speaking, the probability of occurrence of an event is only one element of risk. The consequence of the event is just as important in evaluating risk.

The word *risk* has a negative connotation; you do not often hear of the risk of winning the jackpot, while you may run the risk of failing an examination. In the previous sentence, you would have noted that we used the word *risk* in the sense of chance or probability.

Risk has two aspects. The quantitative (or normative) aspect can be calculated if we know the probability and consequence of an event. The qualitative (or descriptive) aspect relates to people's perception and depends on the emotional state and feelings. Both aspects of risk are important, but their relative importance can differ from case to case. Engineers, physical scientists, and mathematically-oriented people tend to have a bias towards the quantitative aspects. Psychologists and the lay public are more likely to emphasize the qualitative aspects. We need to understand the process by which the customers make decisions; therefore, their orientation or attitudes have a bearing on this matter. If you are to sell your point of view, you must prepare and present your case to suit the target audience's perceptions and decision-making rules.

7.2 DESCRIPTIVE OR QUALITATIVE RISK

People make decisions, consciously or not, on their evaluation of risks. Perceptions play a large part in this process. It will be useful to explore some of the factors that influence it.

7.2.1 Framing effects

A number of factors influence the perception of risk. People exhibit a **risk averse** attitude, when they see the end objective as a gain, and show a marked preference towards a sure smaller gain to a less probable but larger gain.Faced with a situation where there is a 50% chance of gaining $100, or a sure gain of $50, most people will go for the second option. We can compute the expected or risked value of the gain by multiplying the probability of the gain by its numerical value, and this is the same in both cases. If we now reduce the value of the $50 option to say, $45 or even $40, will the decision change? Field experiments show that while a few people will change their minds, most people will still go for the sure gain. True gamblers do not depend on just one deal, but maximize their winnings over many deals. As long as their overall winnings are greater than their losses, they are safe. On average, they will always gain by maximizing the risked gain. The $100 option has a better risked value once the alternative falls below $50, so their decision will change accordingly.

People exhibit the opposite phenomenon when the object is the avoidance of a loss. Here they tend to be **risk seeking**. If there is a 50% chance of losing $100 against a sure loss of $50, most people will opt for the first option. Here too, the expected or risked value of the loss is the same. In this case, they prefer probable high loss to a sure low loss. As before, we can check the sensitivity of the decisions to changes by reducing the value of the sure loss to say, $45 or $40. A few will change their minds, but the majority of people will still prefer the $100 option.

These examples illustrate the so-called framing effect[1]. Researchers have obtained similar results in experiments where the loss is in terms of human lives[2]. Depending on how the researcher described the outcome of the treatment, either as mortality (probability of death), or as survival rates (probability of living), even experienced physicians made significantly different decisions. They also exhibit the same risk seeking or risk averse choices as discussed earlier in the case of gambling[3].

When people look at outcomes, losses appear larger than corresponding gains[4]. Consider the following proposition. You have $1000 to invest, and there is an immediate opportunity available. Shares in a new software company are on offer, but as you know, these are volatile. The prices may double or halve within a short time from floatation, depending on how well the market perceives the product release. So your investment will be worth $2000 or $500, depending on the outcome. If the experts estimate that chance of either event taking place is 50%, how do you decide? If we offer the choice to a sample of the population, many of them are likely to reject the opportunity altogether. As a risked value, the gain is twice as good as the loss, but the potential loss seems much larger than the gain. The opposite phenomenon occurs when the investment is tiny and the potential gains are enormous. These gains appear worthwhile even when the chances of winning them are negligible, or even close to zero. Lotteries operate on this principle, taking small sums from millions of people, and giving a few winners very large prizes. The lottery operator can never lose, as long as there are sufficient buyers of tickets with the dream of winning the jackpot. In this case, the investment appears smaller than it is, while the size of the prize hides the fact that the probability of winning it is negligible. Statistically, no single individual has any reasonable chance of winning the jackpot. However, all the players believe that these chances apply to others, and not themselves.

7.2.2 The influence of choice

The addition of choice can alter decisions, and may reinforce or reject the earlier selections. Thus an additional option can lead to a deferral of the decision, or in other cases, more emphatically confirm the earlier decision. Tversky and Shafir[5] conducted a number of field experiments to examine this effect. The experiments are along the following lines.

You want to buy an item, but have not chosen the model or make. There is a sale of one model at a large discount, in a downtown shop. If the item on offer meets your important requirements, in most cases you will decide to buy it quite quickly. However, if a costlier alternative is available, and perceived to be good value for money, you are likely to wait to gather more data before deciding at all. If the customers perceive the alternative to be inferior, it reinforces their earlier decision.In many cases, the addition of choice simply delays the decision making. It is not because the decision makers have not chosen to do so, but because they have chosen not to do so **now**. Tversky and Shafir[5] have discussed the descriptive aspects of risk in relation to decision making in detail, and they provide experimental evidence in support of their arguments.

Redelmeier and Shafir[6] note a similar situation, in an experiment with a group of physicians. They presented the physicians with a hypothetical case of a patient with a certain painful hip condition. They asked half the group of physicians what they would do when one effective drug was available to relieve the pain. In this case, 47 percent of the physicians in the set elected to prescribe the drug. They asked the other half of the group the same question, but this time another equally effective drug was available to relieve the pain. As an unbiased observer, one might expect that all or at least a large percentage of the second set of physicians would prescribe one of the two drugs. They now had an alternative if they did not favor the first drug. However, only 28 percent of the second set of physicians decided to administer any drug at all. The reality is the exact opposite of the expectation, and demonstrates the influence of choice on decision making. *Clearly, choice itself is a critical parameter, and has a strong influence on the timing of decisions, often resulting in their being postponed.*

7.2.3 Control of situation

The Department of Transport in the U.K. published the following statistics[7] for the period 1986-1990. The figures show the number of deaths per 1,000,000,000 kilometers traveled.

passengers on scheduled UK airlines	0.23
railway passengers	1.1
car and taxi drivers and passengers	4.4
two-wheeled motor vehicle drivers/passengers	104

These statistics indicate that driving is an order of magnitude more hazardous than flying. Yet many of us would not hesitate to drive a long distance, even when flying is a viable option. There is almost an implicit faith that the statistics of bad outcomes relate to other people and not ourselves. As a further example, note that we would be willing to take risks with our own lives engaging in activities, such as para-gliding or bungee jumping. If our children wished to engage in these activities, we may be less comfortable.

People do not decide using facts alone, and numbers by themselves do not always convince them. They know that the statistics presented are often not relevant, so they do not pay attention to them. Just because the vast majority of people die in bed, it does not mean they should not go to bed!

7.2.4 Delayed effects on health

Fear of the unknown has a strong influence on decisions, especially in matters that may have a delayed or long-term effect on health. The introduction of the drug Thalidomide for use by pregnant women resulted in the birth of many deformed and disabled children. Many were born without one or more limbs. This caused a lasting dread of all drugs, and distrust in the people who released them. The public now requires a much higher level of proof from the scientific community before they are willing to accept any new drugs. Drug companies now have to carry out extensive trials before they release new drugs.

7.2.5 Voluntary risks

People willingly accept high risk activities such as smoking or rock-climbing, since they make the decision themselves freely. Participating in high-risk sports is another example of voluntary action. The pleasure that such activities bring is apparently adequate compensation for the potential pain they may bring. If people believe that others are imposing the risk on them, this can prove unacceptable. Health risks at the workplace or in public places such as airport terminal buildings fall in this category. When people see work as a chore or something to endure, rather than a pleasurable activity, they object to these imposed risks. There may be a lesson here; if we see work as a pleasurable activity, more of the risks may become acceptable.

7.2.6 Risks posed by natural phenomena

A single volcanic eruption or forest fire may cause significant pollution with respect to greenhouse gases. The forest fires in Indonesia in the summer of 1997 darkened the skies in Malaysia, Singapore, and Indonesia for days on end. People suffered severe health problems throughout the region. One large plane crashed in Indonesia as a result of the smoke and poor visibility. Human activity initiated the fires, but it was the lack of seasonal rain that caused their rapid spread. In turn, they blamed El Niño for the change in weather patterns. As a result, they treated the whole sequence of events as a natural disaster. The public takes such natural events in stride, even though the effects may be one or more orders of magnitude greater than say, the emissions from industrial activity. As a second example, consider the effect of radioactive emissions. Granite houses can sometimes have radio activity levels much higher than the natural or background level. Thus, people living near a nuclear reprocessing facility may have a lower exposure to radiation than those who live in granite houses. Yet the former feel far more exposed than the latter. Newspapers can improve their circulation with a story about radio active leaks from a nuclear reprocessing facility. It is difficult to raise the circulation with a story on the hazards of living in granite houses!

7.2.7 Subjectivity

In the United States, the American League of Women Voters, as well as college students, rate nuclear power as the number 1 risk to society. Experts rate it way below, at number 20. Motor vehicles rank number 1 with experts, number 2 with the women voters, but only number 5 with college students. Police work ranks a high 17 with experts, but the other two groups think it is only the 8th worst. The table[8] of risk perceptions makes fascinating reading. It illustrates how our personal beliefs or bias affects our perception of risk.

7.2.8 Morality

Fatalities associated with vehicle accidents are much more than deaths due to murders. Should the police concentrate on dangerous drivers or in nabbing suspected murderers? The first course would probably save more lives, but this policy would be socially unacceptable. The fact that one set of deaths is not intentional reduces their emotional severity. Issues of morality come into play in different ways, and influence the way we deal with them.

7.2.9 Dreaded consequences

The public outcry in Europe over bovine spongiform encephalopathy (BSE), or mad-cow disease in Great Britain in 1996-97, had a lot to do with its possible link to the human equivalent CJD or Creutzfeldt-Jakob disease. The main driver was that CJD had no known cure. Heart disease kills many more people than cancer, but usually it does not expose the patient to as much suffering. If detected in time, one can deal with heart disease and, in many cases, limit the damage. One cannot detect some cancers in their early stages. Beyond a certain stage, these cancers are terminal.

The public believed that nuclear power was extremely dangerous, as a result of the Three Mile Island nuclear power plant incident in Harrisburg, Pennsylvania, in 1979. People in the USA began to expect a doomsday scenario with nuclear power. The Chernobyl disaster took place in 1986, resulting in the death of about 300 people, and the contamination of over a million people. This further reinforced their fears, and the industry is in serious difficulty in the USA.

7.3 FACTORS INFLUENCING DECISION-MAKING

These are some of the reasons why people decide the way they do. Our decisions may appear illogical to others who have a different set of values. The underlying reasoning does not follow a simple structure, and so conventional logical analysis is not always the answer. There is no simple right or wrong way, and it is important that we understand that such a decision-making process is normal. The most rational and logical amongst us still decides under the influence of some of these factors. We may, for example, still buy a car based on the smell of the seat leather. When people fall in love, do they use logic to decide?

When we encounter resistance to change from those who will benefit from a reduction in their own risk, we may conclude that they are illogical. The reality lies in our own poor marketing technique—our reasoning may not have appealed to the perceptions of the people involved. Implementation of change needs careful consideration of perceptions, or it will not succeed.

Slovic[8] and his team explained why people resist change, using their factor space theory. Dread and fear of the unknown are two of these factors that rank high in their evaluation. Using these two factors along the X and Y axes respectively, they plotted the response of people to questions relating to about 90 hazards. These included, for example, sporting and recreational activities, household appliances, hallucinogenic drugs, medicinal drugs, DNA research, nuclear power, satellites, nerve gas, solar power, and jumbo jets. Surprisingly, a number of relatively hazardous sports such as mountain climbing and down-hill skiing rank low along both axes. Nuclear power and nerve gas rank high on the dread scale, but the former also ranks high on the unknown risk scale. The injury and fatality statistics may not match these perceptions, but this is how the participants in the study perceived these risks. Whether we agree with them or not, people will continue to make decisions based on such perceptions, and no amount of statistics will help change their mind. *If we are selling a service or product, or trying to persuade people to behave differently, we must remember that our story line must appeal to their perceptions. Unless we do so, there will be no sale!*

7.4 THE QUANTITATIVE ASPECTS OF RISK

Let us now examine the second aspect of risk. This is its quantitative aspect, which we define as follows.

Risk = Probability x Consequence, or alternately,
Risk = Frequency x Severity

We calculate the risk using the estimated or measured value of the two parameters in the equation, as there is no absolute measure. The units are in terms of money, loss of life, ecological or environmental damage.

7.4.1 Failure

Failure is the inability of a process plant, system, or equipment to function as desired. Thus, when there is a failure, we cannot produce widgets or serve customers. Similarly, when traffic jams take place, there is a system failure. In other words, performance will drop to a level below predetermined acceptance standards. Every process is a susceptible to failure. In Chapter 4, we observed that minor failures can be quite useful, because they give us the method to control the process.

If the minor failures occur very frequently however, there is a chance that some of them may escalate to a higher level. Thus, if you have a high frequency of small fires in an installation, there is a distinct possibility that one of them will escalate into a major fire or explosion. Similarly, in an installation that experiences many minor injuries, one can expect a lost-time injury sooner or later.

7.4.2 Exposure

Let us now examine the concept of exposure. If you have to cross a road frequently, your exposure to a road accident is higher than if you did not have to cross the road. The traffic density also affects the exposure, rising as the traffic increases. The demand rate, that is, the number of times we call on something to work, is the industrial equivalent of exposure. Thus a pressure relief valve (PRV) operating close to its set pressure will have a higher demand than an identical one whose set pressure is well above its operating pressure. If there is a wide fluctuation in the operating pressure, there will be a greater demand on the PRV to come into action. These are illustrated in Figures 7.1 and 7.2.

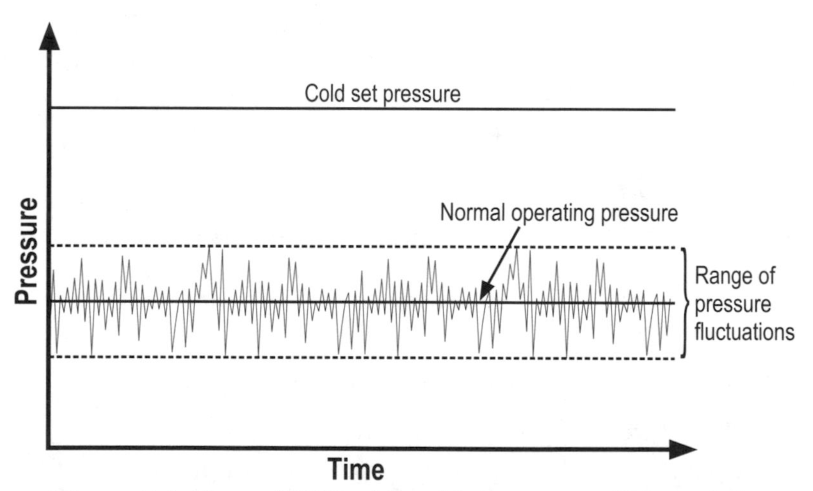

Figure 7.1 Chart of PRV with cold set pressure being much higher than normal operating pressure.

Steady state or 'tram-line' operations have a low demand or exposure compared to processes that experience wide swings. If the process parameters fluctuate considerably, it is less predictable. In many cases, the demand rate may be outside our control and we can only react to the situation. If the demand rate is within our control, for example, the acidity or pH of a chemical process stream, it would be prudent to address this parameter first.

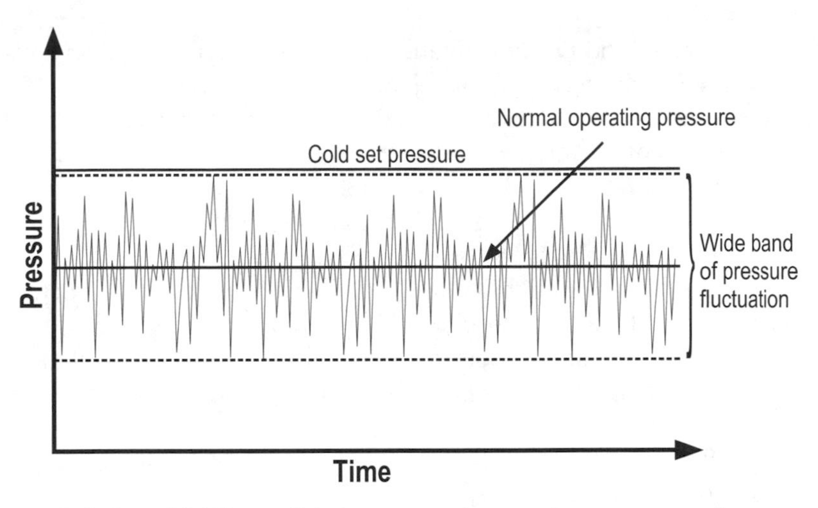

Figure 7.2 As with Figure 7.1, but normal operating pressure close to cold set pressure; with also a wider band of pressure fluctuations.

7.5 CHAPTER SUMMARY

The word risk can have different meanings in English, depending on the context. Quite often, it means chance or probability. Perceptions of risk are important, so we examined the relevant issues. Whether the end objective is a loss or gain affects our attitudes: risk seeking or risk averse. The addition of choice often delays the decision-making. Our bias depends on a number of factors: whether we are in control of the situation, whether they result in delayed effects on health, or whether the cause is natural or man-made. Morality, dread, and subjectivity also influence our attitudes. The important point to note is that perceptions affect decision-making. When we seek the support of an individual or a group, it is as important to appeal to their perceptions as to the hard facts.

We discussed the quantitative aspects of risk, starting with the definition of risk. We examined the salient points of failures and how infrequent minor failures can actually help control the process. However, if their frequency is high, there is a possibility that one of them will escalate into a major incident.

Finally, we looked at exposure or demand rate. Using examples, we tried to understand the impact of a high demand rate. We also examined the advantage of having a process with a low demand rate, or the so-called tram-line operation. We noted that lowering the demand rate to the extent possible is the first step to take in reducing the risks.

References

1) Tversky, A., and D. Kahneman. 1986. "Rational Choice and the Framing of Decisions." *Journal of Business* 59: 251-278.

2) Tversky, A., and D. Kahneman. 1981. "The Framing of Decisions and the Psychology of Choice." *Science* 211: 453-458.
3) McNeil, B.J., S.C. Pauker, H.C. Sox, and A.Tversky. 1982. "On the Elicitation of Preferences for Alternative Therapies." *New England Journal of Medicine* 306: 1259-1262.
4) Tversky, A., and E. Shafir. 1992. "The Disjunction Effect in Choice Under Uncertainty." *Psychological Science* 3: 305-309.
5) Tversky, A., and E. Shafir. 1992. "Choice under conflict: The Dynamics of Deferred Decision." *Psychological Science* 3: 358-361.
6) Redelmeier, D., and E. Shafir. 1995. "Medical Decision Making in Situations that Offer Multiple Alternatives." *Journal of the American Medical Association* 273: 302-305.
7) *Risk, Analysis, Perception and Management*. 1992. The Royal Society. ISBN: 0854034676. 79.
8) Slovic, P., B. Fischoff, and S. Lichenstein. 1980. "Facts and Fears: Understanding Perceived Risk." *Societal Risk Assessment: How Safe is Safe Enough?* Plenum Press. 190-191.

CHAPTER 8

The Escalation of Events

What is it that scheduled airlines do that allows us to take a commercial flight without worrying about our personal safety? How do some industries processing hazardous materials consistently report good safety results? Is it safer to work in some firms than in others?

In this chapter we will trace the events leading to a number of well-known disasters that had taken place in industrial plants or public services during the last few decades. We have chosen to examine the Piper Alpha offshore platform disaster at some length, as it has many lessons to offer. We will also study some other disasters in lesser detail. A common pattern emerges from these reviews. We can see the role of people, plant, and procedures and how they might have prevented the escalation of minor events into major incidents. We will develop a model to help understand the reasons for event escalation and hence how best to prevent disasters.

Disaster inquiry reports usually highlight one or more of the following areas of concern. You will be able to identify these elements as you go through the narrative describing the selected disasters.

- lack of or poor management systems
- poor design
- poor communications
- inadequate procedures
- poor maintenance
- inadequate training
- time pressure on work force

8.1 LEARNING FROM DISASTERS

Are industrial disasters unavoidable consequences of working, or can we learn to prevent them? If we are to do so successfully, the first step is to understand why they occurred in the first place.

8.1.1 The Challenger space shuttle explosion

On January 28th 1986, the Challenger space shuttle took off at 11:38 a.m., and exploded 73 seconds later, killing all seven astronauts. A Presidential Commission of Inquiry investigated the incident, under the chairmanship of the

Secretary of State, William Rogers. Richard P. Feynman, a Nobel Laureate and a well-known Professor of Physics at the California Institute of Technology at Pasadena, was one of the members of the commission. In his book[1], he explains the progress and outcome of the inquiry. The direct cause of the incident was the loss of resilience of the O-rings in the field joints between the booster rocket stages. However, this was not the first time that hot gas had leaked past these joints. The Morton Thiokol Co., which had designed the seal, had analyzed its performance during every previous launch. In one of their studies, they had correlated the seal failures with the ambient temperature at the time of launch. They had a theory as to why the blow-by or leak occurred. The low ambient temperatures resulted in loss of resilience of the seal, and this could explain the incidents. On the night before the disaster, they warned NASA not to fly if the ambient temperature was less than 53°F. NASA was under tremendous political and media pressure not to delay the launch, and the negotiations between them and Morton Thiokol carried on late into the night. The managers of Morton Thiokol and NASA decided to proceed with the launch, in spite of the scientific advice to the contrary. Feynman concludes that there was a failure in management in NASA. Had their controls been effective, they would have learned from previous near-misses.

8.1.2 The Piper Alpha Explosion

This was the worst disaster in the offshore oil and gas industry, resulting in the death of 169 people. On the evening of July 6, 1988, there was an explosion and fire on the Piper Alpha platform. The blast and fire were so severe that two-thirds of the structure collapsed into the sea, and 167 of the 226 people on board, and 2 from a fast rescue craft died. The Court of Inquiry[2] conducted by Lord W.G.Cullen had to reconstruct the events leading to the disaster from the accounts given by the survivors, witnesses on the support vessel and others in the vicinity. Most of those involved directly perished in the disaster, so this task was not easy. *Tharos*, a semi-submersible vessel was anchored about 550m west of Piper Alpha. Purely by chance, an off-duty mobile diving-vessel pilot on board *Tharos* was getting ready to take some pictures of the platform for his child's school project, when the first explosion occurred. He continued to take photographs as the event escalated. A technician on *Lowland Cavalier*, a standby-vessel, also took some photographs. These photographs proved to be valuable in piecing the evidence together.

Piper Alpha was off the coast of Scotland, 110 miles north-east of Aberdeen, in the North Sea. It had pipeline connections to three other platforms, Claymore, Tartan and MCP-01. Piper Alpha supplied gas to Claymore, as the latter did not produce enough gas to run its own gas turbines. Gas export from the Tartan platform line was through Piper Alpha. The combined gas export was through MCP-01 to St.Fergus on the north-east coast of Scotland, about 110 miles away. The oil export lines from Piper Alpha and Claymore merged into a single line to the Orkney Isles, about 128 miles to the West.

Prior to July 6, major construction work was in progress. This included welding work, normally allowed till 2100 hours. The production records showed that the water content in the oil was high, at 10% against the normal level of 2%. The removal of oil from the produced water was by hydro-cyclones, and the clean water discharged to sea. The high water content resulted in overloading of the hydro-cyclones. In order to facilitate the major construction work, they made modifications to the dump-line from the hydro-cyclones to the sea. The result of all this was that some hydrocarbons remained in the discharged produced water. On the evening of July 6, they reported that this discharged produced water was bubbling, evidently due to entrained gas. Numerous gas alarms had been recorded. These could initiate the automatic fire water deluge system. As welding work was in progress at the upper level, they switched off the automatic deluge system.

External communication with Aberdeen was through a tropospheric scatter system.There was a line-of-sight microwave radio link to Claymore, Tartan, and MCP-01. There was a tropospheric connection from MCP-01 to Aberdeen, but not from Claymore or Tartan. On July 6, the direct link from Piper Alpha to Aberdeen was down for servicing.

The supply of water for fighting fire was from two utility/fire pumps, one of which was electric-motor driven and the other diesel-engine driven. There was a dedicated diesel-engine driven fire pump as well. Normally, the two diesel-engine driven pumps were on manual control whenever diving in the vicinity of the suction pipes of the pumps was in progress. When this was so, they had to start the pumps from the local panel, and not from the control room. In the event of a major emergency, the operators would have difficulty reaching the diesel-engine driven pumps, if the fire was in the way.

In order to prevent the formation of hydrates (crystalline ice-like solids) in the colder parts of the process, they injected methanol at various points.The *Joule-Thomson(J-T)* gas-expansion valves and the downstream flash drum were the coldest parts, where hydrates formed easily. The hydrates could cause blockage of the centrifugal condensate booster pumps and then the reciprocating condensate injection pumps (G-200 A & B). This would cause a trip of the pump(s), possibly accompanied by some internal damage. As long as one booster and one injection pump were working, the process would continue to operate. If not, the rise of liquid level in the flash drum would cause a process trip. In March 1988, an internal report stated that the methanol injection rates were lower than required, and proposed additional injection capacity. The situation became worse when any of the methanol injection pumps was down for planned or unplanned maintenance. These pumps were not very reliable, and had frequent long duration breakdowns. On July 6, 1988, one pump was shutdown at 1600 hours and restarted at 2000 hours.An expert later estimated that this four hour interruption would result in the formation of about 250kg of hydrates. The expert estimated that once the injection into the *J-T* valve restarted, the hydrates would break off the walls of the flash drum. They would then move through the booster pump and block the inlet pipe of the condensate injection pump by about 2145 hours. This explanation is consistent with the trip of the G-200B pump, which started the chain of events. Figure 8.1 shows a simplified process flow diagram of this part of the plant.

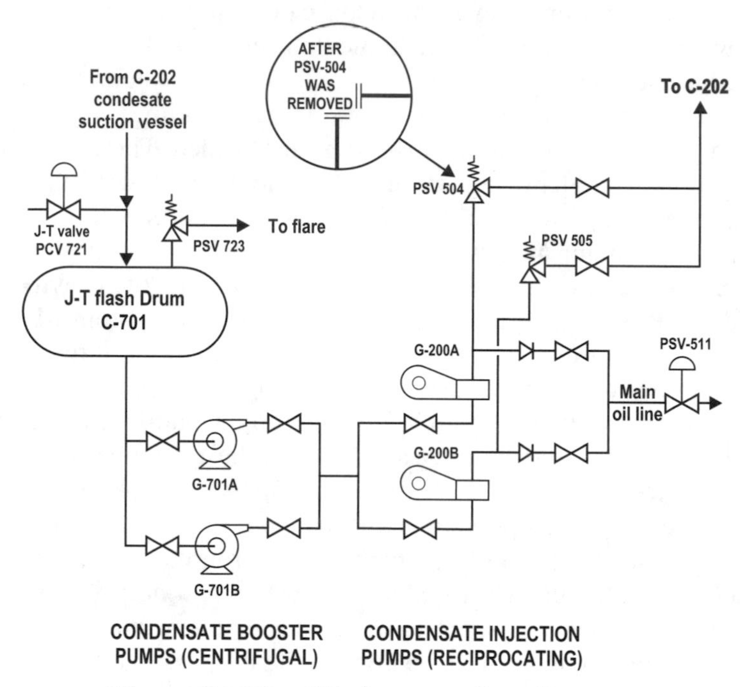

Figure 8.1 Simplified process flow diagram.

During the day, the condensate injection pump G-200A was isolated for scheduled maintenance. The permit to work (PTW) indicated the required electrical and process isolations. Around this time, a program of routine re-certification of pressure safety valves (PSVs) was in progress. PSV-504, located on the condensate injection pump G-200A, was due and hence removed for this purpose under a separate Permit to Work (PTW). The PTW for the PSV-504 did not refer to the PTW for the pump G-200A maintenance and vice-versa. If the operator saw only one of them, there was no way to know automatically that some other work was also in progress. By about 1800, hours PSV-504 was ready for refitting, but at that time the crane was not free, so they postponed the work to the next day. The fitter working on the PSV was aware that scheduled maintenance was in progress on the pump itself, so it would be reasonable for him to believe that it would be down for some time. He installed blind flanges on the open ends of the pipes. In his mind, their purpose was to stop foreign matter entering the piping, and not for containing fluids under pressure. So the bolts were probably only hand-tight and not flogged, as he would have done normally. The suspended PTW for PSV-504 was not in the control room, as required by the procedure. Around 2150 hours, condensate injection pump G-200B tripped and could not be re-started. The operators did not realize that the PSV-504 was not in place on pump G-200A. They assumed that the pump alone was under normal scheduled maintenance. In the hurry to start pump G-200A, they located the pump isolation permit, and re-connected the pump electrically. While all this was going on, both pumps G-200A and B were out of commission. The upstream vessel liquid level rose, tripping the reciprocating compressors. A set of gas alarms came on in rapid succession before the first explosion took place.

Subsequent expert evidence and wind tunnel tests established that the size of the first explosion required about 45 kg of fuel. After considering several leak scenarios, the Court of Inquiry concluded that the blind flange joint on the discharge pipe of G-200A pump leaked, when they pressurized it for start-up. On a balance of probabilities, the Court believed it was the most likely scenario.

A fatality occurred earlier, on September 7, 1987. A rigger died due to a fall, and the remedial actions by the company included instructions to the PTW issuing staff to state the full scope of work clearly. There was evidence to show that the workers violated these instructions routinely. The company did not enforce the procedures. Clearly, there were weaknesses in implementing the company's own PTW system. Another weakness was the poor hand-over from the day-shift to the night-shift. These two weaknesses surfaced again on July 6, 1988, with disastrous consequences.

In a major emergency situation, the fire water requirements were such that they needed the diesel fire pumps to supplement the electric fire pumps. Remote starts of the diesel fire pumps from the control room were not possible, once they were in the manual control mode. Local panel starts were the only available option. In the summer months, when there was a lot of diving work, the practice on Piper Alpha was to leave the diesel fire pumps in the manual mode from 1800 to 0600 hours. In June 1983, an internal fire protection and safety audit report recommended that these pumps be kept in the automatic mode as long as there was no diving work near the pump intakes. However, the offshore installation managers (OIMs) continued the practice of setting them to manual whenever **any** diving work was in progress, irrespective of its location. As a result, on July 6 the diesel fire pumps were in the manual mode. In this condition, the fire-water system capacity was inadequate to tackle a major emergency.

On Piper Alpha, they routinely tested the fire-water deluge system every quarter. In May 1988, during such a test, they found blockage in about 50% of the spray nozzles. They ordered replacement pipe work on a high priority, and planned to complete it in June 1988. In the event, they could not complete this work in time. However, this was not the first time they observed blocked nozzles. In the February 1988 tests, they found several blockages. As early as 1984, they had recognized deluge pipe work and nozzle failures. They initiated replacement actions in June 1986, but delays in design and construction meant that progress was very slow. Important parts of the platform continued to have poor deluge systems. However, the ship surveyor from the Department of Transport did not find these defects during the biennial inspection. Thus the regulator's inspection was ineffective.

Many of the survivors stated that they had not received a safety induction course, and some others said that it was brief and cursory. They had not carried out evacuation drills at the stated frequency. In the preceding three years, they had not practiced full-scale emergency scenarios. Similarly, staff on specialist duties did not practice weekly drills in six special subjects including fire-fighting. All staff working on offshore installations had to undergo a combined fire-fighting and survival course, at the end of which they received a certificate. It was up to the company to verify that their own staff as well as their contractors' staff held valid certificates. After the accident, the police found that as many as 21 of the deceased did not hold such certificates.

Both Claymore and Tartan were aware of a major emergency on Piper Alpha, but continued production, resulting in large flows of hydrocarbons that fed the fire in Piper Alpha. Even after the rupture of the Tartan riser at 2220 hours, which event was clearly visible from Claymore, it continued to operate at full capacity. These actions contributed to the rapid escalation of the fire on Piper Alpha. The Court of Inquiry concluded that the training of the three OIMs did not help them to deal with such a scenario. They were not ready to deal with an emergency in which an explosion on one of the platforms put it out of commission. The lines of communication were clearly inadequate and responses too slow in the face of the emergency situation.

The Court of Inquiry made a number of observations about the events leading to the disaster. The following is a partial list:

- The operating staff had no commitment to follow the written procedure of the PTW system; the people knowingly disregarded the procedure. The night shift treated the extension of canceled PTWs casually.
- The PTW depended on informal communication; this failed to prevent the night shift from re-commissioning the condensate injection pump G-200A on July 6, 1988.
- They did not provide adequate and effective training on the use of the PTW system.
- The hand-over at the end of shift was deficient; this was demonstrated both on September 7, 1987, when a fatality occurred, and later on July 6, 1988.

- They could not start the diesel fire pumps from the control room when they were in the manual mode.
- Regardless of the location where divers were working, the diesel fire pumps were in the manual mode, and out of service for extended periods. This practice contravened the internal audit recommendation.
- They knew that the fire-water deluge system was in a very poor condition over a period of four years, and that there were many delays in replacing the piping. The defective system was still in place on July 6, 1988.
- There was no structure or format to safety induction courses. These were casual and informal sessions and sometimes not given at all.
- They did not organize emergency drills, evacuation exercises and training in emergency duties at the required frequency.
- They did not check on-shore training certificates in fire-fighting and survival courses properly.
- The company had a proper safety system on paper, but the quality of management of safety was ineffective.

8.1.3 King's Cross underground station fire

A fire started in the London Underground Kings Cross station on November 18, 1987, at 7:25 p.m. In all probability, it started in a pile of rubbish, under the track of an escalator. The tracks of the escalator were wooden and may explain its rapid spread. The authorities took prompt action to limit the damage when they realized the scale of the fire. They ordered the incoming trains not to stop at the station, so as to minimize the number of people exposed to the fire. However, the train drivers did not receive the instructions, and continued to stop at King's Cross, allowing people to disembark. There was no evacuation plan in place. With many exits closed, the fire and smoke spread, and resulted in the death of 31 people.

8.1.4 Milford-Haven refinery explosion

During a severe electrical storm in July 1994, lightning struck the refinery, resulting in plant upsets. As a result, there was a release of about 20 tons of hydrocarbons from a flare knock-out drum. This formed a vapor cloud which ignited about 110 meters away and exploded. A combination of events contributed to the disaster. For example, the control panel graphics did not provide a proper overview, and a closed control valve appeared on the panel as if open. Also, a completed plant modification did not have a supporting risk analysis. The Health and Safety Executive (HSE) who carried out the investigation[3] concluded that there was a combination of failures of management, control systems and equipment.

One of their recommendations was to reduce the number of instrument trip and alarm functions to match the risk levels.

8.1.5 Bhopal

On December 3, 1984, there was a leak of methyl isocyanate from a storage tank at a chemical plant in Bhopal, India. This resulted in a vapor cloud engulfing the surrounding shanty town. About 2500 people died, and some 25,000 people suffered injuries. This was the worst disaster in the history of the chemical industry.

A load of methyl isocyanate arrived in the plant for use in the process. The operators believed that they were loading it into a dry tank, but this was an incorrect assumption. The water caused a violent reaction and the relief valve on the tank lifted. The vapors from the relief valve should have gone through a scrubber designed to absorb them. A refrigerant cooling system should have kept the tank cool, thereby reducing the intensity of the reaction. Both the scrubbing system and the cooling system were out of commission[4], resulting in the disaster.

8.1.6 Chernobyl

On April 26, 1986, unit 4 of the Chernobyl nuclear power station experienced a sudden surge of power at 0124 hours. This surge of between 7 and 100 times normal operating power happened within approximately 4 seconds. The safety systems could not respond in time, causing rapid coolant vaporization and resulting in a catastrophic steam explosion. The reactor top was blown off, and this exposed the core to air. This caused a hydrogen explosion, which led to the graphite moderator catching fire.The uranium fuel particles escaped along with the gases from the fire. The radioactive debris covered large parts of Ukraine, Byelorussia, Russia, Poland, Lithuania, Latvia, Estonia, Finland, and Sweden.

The Chernobyl Commission Report[5] gives the following sequence of events. The authorities planned an experiment to evaluate a modification of the turbo-alternator to generate power when it was coasting down. This was timed to coincide with a scheduled reactor shutdown. The first event occurred about 24 hours earlier at 0100 hours on April 25 when they began to reduce reactor power to the 50% level. This took about 12 hours, and they switched off one of the two turbo-alternators. Shortly thereafter, at 1400 hours operators turned off the emergency cooling system, as it would interfere with the experiment. At 2310 hours, they started reducing reactor power to the 25% level. For this purpose, they had to switch from the local automatic control to the global automatic control. In the local control case there were sensors located inside the reactor core, while in the global control case, they were on the periphery of the core. This switching operation was done at 0028 hours, but due to an error, the power level dropped to less than 1%. This led to xenon poisoning of the reactor, so the operators raised the power level again. After half an hour, the power was back up to about 7%, but to do this they had pulled out all except six control rods. At this point the reactor was unstable, and any increase in power could cause a rise in output. At about 0122 hours they manipulated the water flows to increase the cooling. Due to a slight fall in flow, the controls dropped automatically, and by 0122 hours, things seemed to be back in control. At this point, they took the next step in the experiment, namely to trip the turbo-alternator. This had so far been a good heat sink, and its removal from service initiated the rise in reactor power. At 0124 hours, the reactor became unstable, and instantly reached criticality. The explosion and release of radioactive material resulted in the death of more than 300 people and injury to over one million others. The fatality estimates by some sources are much higher. For example, the New York Times of April 23, 1995 estimates it at 5000 fatalities. Vast areas of the surrounding farm land were contaminated[6].

8.2 HINDSIGHT IS 20-20 VISION

In all of these incidents there is a pattern of some common elements contributing to the disasters. One or more links in the chain have been weak, resulting in an escalation of the event.

In the Challenger case, the less-than-ideal field joints between the booster stages had a blow-by, initiating the disaster sequence. However, this part of the design was weak, and all the concerned people knew this fact. There had been several incidents before this disaster, where a blow-by had taken place. To initiate a blow-by, it was also necessary to have a low ambient temperature. The contractor warned NASA of this situation the night before the disaster. With the help of hindsight, we can conclude that they did not heed the warning, perhaps because of the intense pressure on the people concerned. A good management system could have overcome the political and media pressures, for example, by publishing the results of risk analysis studies. This may have helped to obtain a delay in the launch till such time as the conditions were favorable.

The Piper Alpha Inquiry resulted in far-reaching changes. The management of offshore safety in the U.K. changed significantly, including a change in the regulatory regime. The principal recommendation was the use of a Safety Case regime where it became incumbent on the owner to explain the Safety Management System (SMS) proposed. The SMS had to fulfill three requirements, as follows:

- To demonstrate how it would ensure that the design and operation are safe;
- To identify major hazards and risks to personnel and demonstrate that adequate controls are in place;
- To provide a Temporary Safe Refuge for use by the personnel on board in the event of a major emergency and to provide facilities for personnel evacuation, escape and rescue.

It proposed that the existing prescriptive legislation be replaced by a set of goal-setting regulations. Non-mandatory guidance notes would support these regulations. So as to prevent a conflict of interest at the regulatory level, the Court recommended the enforcement powers of the Department of Energy be transferred to the Health and Safety Executive. There were 106 recommendations in all, divided into 24 subject areas, covering a wide range of topics. These included, for example, legislation, introduction of the Safety Case regime, control of hydrocarbon inventory, fire and explosion protection, emergency procedures, helicopters, drills exercises and evacuations, and training for emergencies.

The King's Cross disaster showed that when large numbers of people are using a public facility, it is difficult to control sources of ignition. As there are many smokers among the users, this task becomes unmanageable. We cannot attribute the King's Cross disaster to the initial fire alone, though it was the obvious starting point in the chain of events. The fact that the escalators had wooden treads increased the speed of propagation of the fire, but we cannot blame even this for the turn of events. The real problem was that the drivers did not receive the instructions from the authorities to drive through and not stop at the station. Lack of an evacuation procedure, and the closure of many exits compounded this matter further.

Electrical storms and lightning strikes are not uncommon, especially in places where there is frequent rain. The design of the plant in a location such as Milford Haven should have taken cognizance of such weather patterns. The HSE report identifies several plant deficiencies, inadequate change control procedures, and a management system that permitted the plant to continue operating under unacceptable conditions.

In the case of Bhopal, plant management failed to regard the unavailability of the scrubber and refrigeration systems seriously. Since entry of water into the methyl-isocyanate storage-tank could result in release of toxic vapors, they should have had safeguards to prevent this eventuality. As the plant handled toxic products, these were serious failures. The government that permitted the growth of a shanty town so close to a plant handling toxic products is clearly culpable. The situation was ripe and ready for a disaster.

The Three Mile Island incident in Harrisburg should have been enough to warn those in charge of the Chernobyl test. A risk analysis of the test procedure would have identified the probability of a runaway reaction. A management system that permitted the high risk test to proceed with all the safety controls defeated is a recipe for disaster. The Chernobyl Commission Report[5] found the direct cause to be the series of errors made by the operators during the experiment. It blames the design of the RBMK 1000 reactor as a fundamental cause. The harshest criticisms were of the Soviet disregard for safety of the plant personnel, local population, environment, and neighboring countries. They blame the direction given by the Soviet Twenty-Seventh Party

Congress of March 1986, which exhorted people to conserve supplies of energy wherever possible. They state that the Chernobyl experiment was a result of this directive.

8.3 FORESIGHT - CAN WE IMPROVE IT?

How can we use the knowledge gained by analyzing past failures to improve future performance? In process terms, it does not matter whether we are manufacturing chocolates, assembling cars, refining hydrocarbons, operating a nuclear power plant, or processing toxic chemicals. From the above discussion, it will be clear that relatively minor events can result in major disasters. In each case, it was possible to stop the escalation, with competent and motivated people, good quality procedures, and the right equipment. A good management system would have ensured the right level and quality of communication. One or more or these links have failed in each of the disasters that we examined.

8.4 EVENT ESCALATION MODEL

At this point, we will introduce a model to explain the process of escalation. Figure 8.2 shows such a model with one level of escalation. At the base of the triangle are the relatively frequent minor failures. These minor failures can escalate into more serious ones.This can take place under certain conditions. The model shows three barriers that could have prevented escalation of minor incidents.

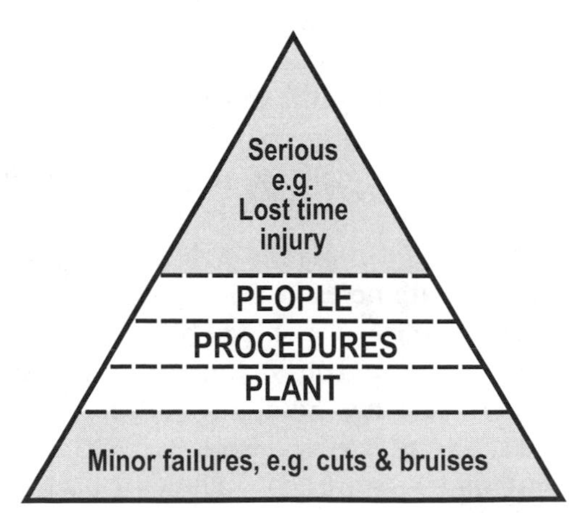

Figure 8.2 Event escalation model.

We use dotted bands to represent these barriers (people, plant, and procedures). We can think of these barriers and the manner in which they work, in the following manner.

People. Competence, training, and motivation enable people to spot and correct the conditions that cause minor failures, and thus reduce their impact. For example, when the dimensions of machined parts approach the limits in the process control chart, the operator replaces the tool tip or resets the machine and brings the process back in control.

Procedures. These are the means of transferring other peoples' knowledge and experience to those operating the process. Typically, manufacturers will tell you how to operate their equipment and software vendors will give you navigation guides and help screens.The knowledge and experience of previous incumbents is the basis of company policies, standards, and procedures. They may have gained some of the knowledge as a result of earlier failures (incident inquiry, customer feedback reports, and audit recommendations). An even wider span of experience forms the basis of statutory instruments, regulations, national laws and international standards.

Plant. The plant consists of the hardware (or software). Designers provide various protective systems to prevent the escalation of minor failures. First-aid boxes, fire extinguishers, smoke, fire or gas detection systems, furnace protection systems, fire extinguishing systems, and emergency shut-down or release systems are all examples of the barriers in this category.

Incident investigation reports will contain some combination of these three ***P***s cited as the reason for the major event. We can trace the escalation to the failure of these barriers, in combination with the fourth ***P***, the process demand rate. The failures of these barriers are unrevealed, or else the conscientious manager would do something about correcting the situation. We discussed hidden failures in Chapter 3, section 3.7, and explained why the availability of the item or system has the same numerical value as the survival probability or reliability. In what follows, we will use reliability and availability interchangeably, noting that it applies only in this special context.

We can visualize the model in a slightly different way, with individual barriers considered as plates with holes in them. A solid plate barrier with no holes would be perfect, or 100% available to block the pellets. A plate with holes has an availability less than 100%. Imagine now that we shoot pellets from below this multiple barrier towards the apex. The holes are large enough to pass a pellet, and each plate is strong enough to stop a pellet.If we shoot many pellets randomly, and there are enough holes in each plate, there will be a few of them in alignment, so that some pellets pass through all the plates. We can visualize event escalation in a similar way. The number of pellets fired represents the demand rate or frequency of minor failures. The pellet or pellets that manage to go past the barriers represent the number or frequency of major events.

Do we require all three barriers each time? If this is the case, we would represent them as a series chain in a reliability block diagram, as shown in Figure 8.3.

Figure 8.3 Series RBD model.

Using Boolean notation, we link the blocks by AND gates. We can calculate the availability of the whole system as the product of the availability of each of the three blocks.

$$A_{system} = A_{people} \times A_{procedures} \times A_{plant} \qquad 8.1$$

where $A_{subscript}$ is the availability of the individual barrier named.

If on the other hand, the barrier would be effective as long as any one of the three ***P***s worked, they would be in parallel, as shown in Figure 8.4.

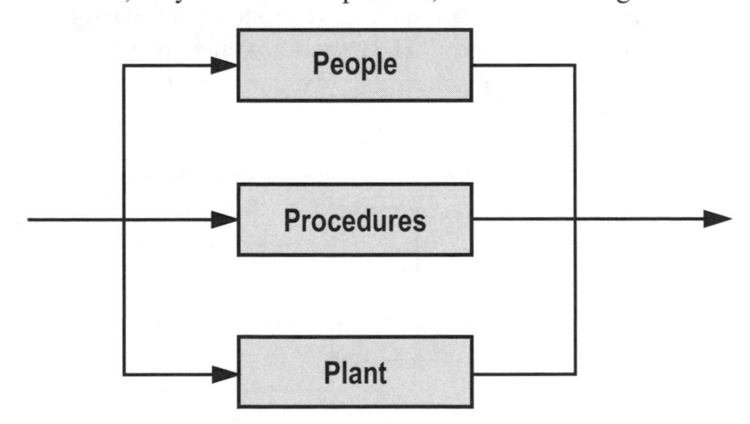

Figure 8.4 Parallel RBD model.

Using Boolean notation, we link the blocks by OR gates. We calculate the system availability using the following expression:

$$(1 - A_{system}) = (1 - A_{people})(1 - A_{procedures})(1 - A_{plant}) \qquad 8.2$$

In most cases, the plant barrier would be a pre-requisite. For example, in the case of a fire, you would need fire fighting equipment such as extinguishers, sprinkler systems, or fire trucks. We can only treat injuries if we have medicines, bandages, and medical facilities. For the purpose of this discussion, all these physical aids fall under the category of plant. The next requirement is people who will use these aids or plant. If the people are competent, trained, and motivated, they know the right procedures to use in each circumstance. In such a case, the need for written procedures is minimal. In most cases, however, it is unlikely that everybody knows exactly what to do or not do, when and whom to communicate with, or even the right sequence to use. In all such cases, we need written check-lists and procedures. Similarly, we can compensate for poorly-trained staff by making good quality procedures available. As an example, think of the situation when you are a hotel guest. You are not familiar with the location of fire alarm stations—the little glass-covered boxes that you have to break to initiate an alarm. Yet all the guests must know how to use them, so the hotel needs a procedure. Further, they display it prominently, as they have to make sure the guests notice the procedure. Hotels do this by displaying the procedure on the inside panel of the main door, at eye level. In this case, the procedures barrier supports the people barrier.

You encounter a different situation when you call in a vendor representative to assist you in carrying out a machine overhaul. In this case, you may have detailed procedures for the dismantling, repair and re-assembly of the item of equipment. However, you may encounter unusual situations, which these procedures do not cover. This is when the expertise and the knowledge of the vendor representative come in handy. The expert has encountered many unusual situations and can improvise a solution to overcome your problem.

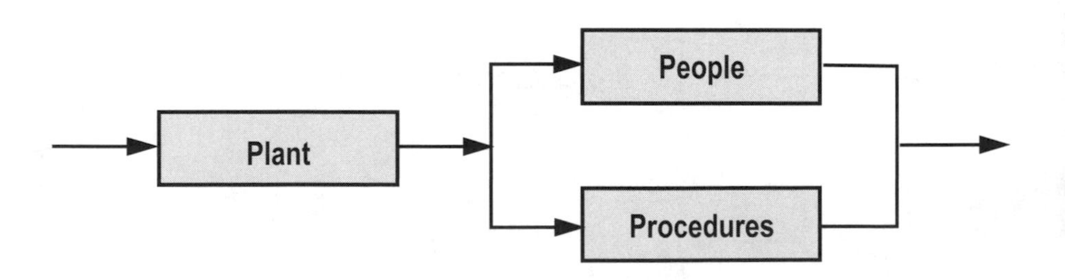

Figure 8.5 Reliability Block Diagram

In this case, the procedures barrier is less than perfect, but the people barrier tends to compensate for the weakness. These examples illustrate the reason why the people and procedures barriers can be considered as alternatives, so that they are in parallel in the reliability block diagram. Figure 8.5 shows the corresponding RBD.

This configuration will change from case to case, but the above reliability block diagram is fairly representative. The next difficulty is that an objective method to measure the reliability of the people or procedure barriers is not available. Quite often though, we can judge the relative or incremental value, and this can be useful. We can estimate the reliability of the people and procedures barriers. If in our judgment, the reliability of the people barrier is low, we should take extra care to ensure that the procedures are of high-quality, and are well understood. The reliability of the people-barrier depends on their training, attitudes, and motivation. The prevailing environment or culture will have an influence on attitudes. The reliability of the procedures depends on those who wrote them, and whether the circumstances are the same today as those that were prevalent when they wrote them. The utility of this model is to assign relative importance and to check the sensitivity of each barrier.

Using this RBD, the system availability is given by

$$A_{system} = A_{plant} \times \{1 - (1 - A_{people})(1 - A_{procedures})\}$$

or,

$$A_{system} = (A_{plant}) \times \{A_{people} + A_{procedures} - (A_{people} \times A_{procedures})\} \qquad 8.3$$

The higher the system availability, the better it is able to cope with event escalation. It follows that the higher the process demand rate, the tighter the barrier should be and the higher the desired availability. An examination of the above expression shows that a high plant availability is an essential prerequisite to meet this objective. Some flexibility is available in the case of the remaining two barriers.

The reality is more complex than illustrated in the model. The barrier availability can change with time. As an example, consider the motivation of people, a factor that can determine how they respond to a given situation. Many factors including emotions and feelings affect motivation. Thus, events such as an argument with your spouse at breakfast, winning a golf match the previous weekend, or the death of Diana, Princess of Wales, can influence your morale and motivation. This is why barrier availability is not a firm and constant number.

Next, take the case when a procedure exists and one has the training to deal with a given situation. At the crucial point, some other event may divert one's attention, or one may simply forget the required procedure. Designers of control panels have to take care to minimize the number of alarms so that operators do not face an information overload. Often, the cause of pilot errors is the need to process large volumes of information very quickly. A period of high stress, whether physical or emotional, can cause loss of concentration. What we often call 'bad luck' is often the low availability of the barriers at a time of high demand.

Lastly, we have represented the three elements as independent variables; this is not strictly correct. The attitudes and motivation of people can affect the availability of the procedures or plant barriers.

8.5 DAMAGE LIMITATION MODEL

We can extend the concept to the next level of escalation. Figure 8.6 shows the damage limitation model, using the same principles.The earlier discussion applies to this model as well, but we modify the role of the three barriers to reflect their new function.These are to prevent (or reduce) fatalities, total loss of production, serious environmental damage, or major loss of assets.

Figure 8.6, Damage limitation model.

The new roles of the barriers are as follows.

People. Competence and training of the personnel in emergency response. In this case, motivation is not an issue!

Procedures. Predetermined emergency response procedures, for example, 'Action in case of a fire' notices in hotel rooms, 'Safety Instructions' card in an aircraft, or building evacuation drill procedures.

Plant. Equipment and facilities especially designed to cope with emergency situations, for example, fire-fighting trucks, lifeboats, ambulances, rescue helicopters, oil-slick booms, underground bomb, or nuclear shelters.

We can find out about the soundness of the barriers only when we call on them to work since their condition is hidden or unrevealed. For example, the operation of a fighter-plane pilot's seat ejection mechanism will not be evident to the pilot, unless he triggers the ejection mechanism.We can test the ejection mechanism some time prior to take-off to check its availability. The point is that we must call on it to work, either by a simulated need or because of a real need.

8.6 FAILURE OF BARRIERS

We discussed the Piper Alpha disaster at some length, and can now attempt to identify those barriers that might have avoided the event escalation, or at least reduced the loss of life.

In terms of the event escalation model, we can identify the following representative barrier failures.

People. Inadequate training in the use of PTW procedure; improper shift hand-over.

Procedures. No cross-referencing of PTWs; continuing high production levels when process conditions were poor (water content 10%, gas in produced water, radiation heat due to high flaring levels, several gas leaks) while a lot of hot work was in progress.

Plant. Crane unavailable to refit PSV-504; methanol pumps undersized; frequent and prolonged outage of methanol pumps.

In terms of the damage limitation model, we can identify the following representative barrier failures.

People. Inadequate training in evacuation and escape due to infrequent emergency drills; lack of survival certificates in 21 cases; lack of commitment to safety at all levels; poor leadership by all three OIMs; a safety culture that permitted continuing production ignoring many warning signs; poor audit by Department of Transport surveyor.

Procedures. Diesel fire-pumps on manual control; poor emergency-scenario planning; delays in shutting down Claymore and Tartan.

Plant. Deluge system unavailable; diesel fire-pumps inaccessible and hence inoperable in an emergency shutdown of process; isolation of hydrocarbon streams not initiated automatically; lack of alternative direct communication with Aberdeen when primary system was down for servicing.

8.7 EVENT ESCALATION RELATIONSHIP

We now postulate a hypothesis to relate the minor event frequency, the barrier availability, and the major event frequency. Earlier, we argued that a plant with many minor incidents was likely to have a high incidence of more serious events. Similarly, we discussed the importance of the barriers that prevent escalation. The following expressions represent these arguments:

$$\textit{Frequency of serious failures} \propto \textit{Frequency of minor failures}$$

and,

$$\textit{Frequency of serious failures} \propto (1 - A_{barriers})$$

thus,

$$\textit{Freq. of serious failures} = k \times \textit{Freq. of minor failures} \times (1 - A_{barriers}) \quad 8.4$$

where **k** is a constant.

We can reduce or eliminate serious failures either by minimizing the minor failures, namely, by reducing the process demand rate, or by increasing the availability of the barriers. Reducing the process demand rate is not always possible, as several factors that are not in our control come into play.

The availability of the barrier depends on its intrinsic reliability, or build-quality. We can, in theory, improve the intrinsic reliability of the plant by carrying out design changes. Similarly, we can train people and thus improve their competence. We can revise procedures to ensure that they are current, applicable, and effective.

The benefits associated with such improvements have to be sufficient to justify the cost. The law of diminishing returns applies to reliability improvements as in other aspects of life. As we make the barriers more reliable, the marginal cost of further improvements rises more steeply. This in turn means that a 100% reliability is not achievable in practice. As noted earlier, our interest is in the system as a whole, not just the three component parts. The sensitivity to cost for the marginal improvements to each barrier will be different, so an opportunity for cost optimization presents itself.

Design changes are not always in our control, as equipment vendors may not be willing to execute them. What do we do in such a case? There is a second method to improve availability. We can do so by altering the barrier test frequency. In Chapter 3, section 3.8, we discussed the relationship between the intrinsic reliability, system availability and test frequency.

8.8 EVALUATING TEST FREQUENCIES

We can use expression 3.13 to evaluate the test frequencies, with the assumption that hidden failures follow the exponential distribution. The approximation permits us to compute the test intervals that will give the required mean availability with relative ease. The limits of applicability discussed in Chapter 3 are important, and expression 3.13 becomes invalid outside these limits. Some examples of how we can use these concepts follow.

- By testing smoke detectors once in six months, we may get a mean availability of 94%. For the given intrinsic reliability of the detectors in the same operating context, we can raise the mean availability to 97%, by reducing the test interval to 3 months.
- On hearing the fire alarm, the emergency procedure requires all the occupants of a building to leave it and assemble in the muster point, usually the parking lot. How often should we conduct an emergency drill or what is the test frequency? The answer depends on how well trained and familiar the occupants are with the emergency procedure, or their intrinsic reliability. If they are a changing population, with a significant number of temporary staff, a high test frequency, say once a month, would be appropriate. On the other hand, if the same people have been using the building for a long time, they will be very familiar with the layout of the passages and stairs. In this case, we can reduce the frequency to, say, once in twelve months. In both cases, the availability of the barriers would be comparable. In a plant shutdown, there will be many newcomers and temporary workers. From a risk-based approach, it is not sufficient to run induction programs alone. We have to test the reliability of the staff by carrying out drills.
- A pressure relief valve operating close to its set pressure is prone to lift frequently, especially if the process fluctuations are high. Obviously, the relief valve must lift whenever called upon to do so. In terms of the above model, the process demand rate is high, so we need to improve the barrier availability. We can do this either by improving the build-quality or intrinsic reliability, or by increasing the test frequency. Generic test intervals are not appropriate from a risk management point of view. If we know the intrinsic reliability of the relief valves in their operating context, the process demand rate, and the required system availability, it is easy to calculate the required test interval. The required system availability depends on the consequence of failure of the relief valve. In practice, it is not possible to assess the intrinsic reliability of a single relief valve, as it is unlikely that it will fail many times. Therefore, we collect failure data from a reasonably large sample of relief valves. With a large sample, we can be more confident in the results. Thus the failure rate itself is generic.

  However, the exposure or demand rate on each relief valve can vary quite widely. Similarly, the consequences of the lifting of a relief valve can also vary. As a result, the risk level differs for each case. The required barrier availability depends on the level of risk. In theory, we should vary the test frequency accordingly. This is often not practical, as access to the relief valves will invariably require a plant shutdown. The test frequency of the relief valves exposed to the highest risk often determines the plant shutdown frequency.

These examples demonstrate that rule-bound test frequencies are unlikely to be suitable in managing risk effectively. We can accept generic frequencies only when they are conservative. These will always be excessively stringent in the lower risk situations, which can be a significant proportion of the total. As a result, more often than not, we will end up leaving money on the table.

In order to manage risk effectively, we propose that we examine each case using the following steps:

- Determine the demand rate, is it high or low?
- Use this to determine the required level of barrier availability;
- Check sensitivity of people, plant, and procedures' barriers for incremental value;
- Choose the combination that gives maximum value per $, in terms of availability;
- Calculate the test interval for each barrier.

8.9 INCIPIENCY PERIOD

We have considered hidden failures so far. For completeness, we will also look at evident failures. As the equipment condition deteriorates, symptoms appear, which we can measure. We monitor, for example, the bearing vibration level, the electrical insulation resistance of a motor or the remaining wall thickness of a pressure vessel. The rate of deterioration in condition can help estimate the time to failure. The incipiency period is the time taken to go from the sound to the failed condition, as discussed in Chapter 4.

In Chapter 10 we will see that the inspection interval cannot exceed half the incipiency interval.

8.10 CHAPTER SUMMARY

We began this chapter by examining a number of well-known disasters that had taken place in industrial plants or public services during the last few years. A common pattern appears to emerge, and some of the weak links become evident. These relate to the reliability, competence, and motivation of people, the quality and suitability of procedures, and the design and upkeep of the plant. A good management system is the best tool to ensure that we can meet these requirements.

With the help of a model, we explained the role of people, plant and procedures in preventing the escalation of minor events into major incidents. We represented these three ***P***s as barriers that prevent escalation of events. Holes in the barriers represented the unavailability of the barriers. Using this representation, the more holes there were in the barriers, the easier it was for the events to escalate. The availability of the **people-**barrier is often dependent on the moods and feelings of those involved. As a result, the barrier availability may change with time. Further, the availability of one barrier may affect that of the others.

The demand rate or exposure represents the frequency at which the process demand occurs. When the demand rate is high, the availability of the barriers also has to be correspondingly high. By matching the barrier availability to the demand rate, we can control the escalation of minor events.

Once a serious event such as an explosion has taken place, the first order or business is to limit damage. We must make every effort to minimize injury or deaths, environmental damage, or serious loss of production capability. We use a damage limitation model to explain this process. The same three ***P***s come into play again, but they have slightly different roles. In this case, the primary requirement is emergency response. The actual process of escalation of a serious incident into a disaster is very similar to the event escalation process.

We introduced a hypothesis to relate event escalation to the barrier availability. This relates the intrinsic reliability, test frequency, and barrier availability. We can also use it to calculate test frequencies that will provide the desired level of availability.

Some practical examples illustrate the application of these principles. From these, it will be clear that the principles are uniformly applicable, and are not specific to any one type of industry.

References

1) Feynman, Richard P. 2001. *What Do You Care What Other People Think?* W.W.Norton & Co. ISBN: 0393320928.
2) Cullen, W.G. 1991. The Public Inquiry into the Piper Alpha Disaster. Department of Energy, HMSO Publication Centre. ASIN: 010113102X.
3) "Poor Management Contributed to Texaco Explosion. 1997. The Safety & Health Practitioner, vol. 15, no. 10. October: 3.
4) Bhopal Methyl ntIsocyanate Incide Investigation Team Report. 1985. Danbury: Union Carbide Corp. March.
5) Chernobyl Commission Report web site. http://www.mwukr.co/chcom.htm. For more information refer to_web site http://www-bcf.usc.edu~meshkati/causes.html.
6) USSR State Committee on the Utilization of Atomic Energy. 1986. The accident at the Chernobyl Nuclear Power Plant and its Consequences. Vienna: IAEA. August. Reference documents at website http://www.thyrolink.com/literature/report1996_2/seite08.html

CHAPTER 9

Maintenance

Maintenance can mean different things to different people. Quite often, senior managers and accountants see maintenance as a cost burden that should be minimized. At the working level, some of us see it as a set of preventive, corrective, or breakdown rectification activities. Some classify it as reactive or proactive work. To stillothers, it means predictive, planned, or unplanned activity. All these are merely the various dimensions of maintenance. They are valid descriptions, but do not address its functional aspects. We prefer to look at the role or function of maintenance and its strategic contribution to the health of a business. In Chapter 8, we examined the role of maintenance in preventing event escalation and how it helps retain the integrity and productive capacity of the facility over its life. This is its strategic role; maintenance helps maximize the profitability of a business over its life.

In this chapter, we will see how appropriate maintenance strategies can help manage risk effectively.

In Chapter 2, we noted that the capability of an item of equipment, syste,m or plant may deteriorate over time, due to fouling, wear, corrosion, or fatigue. At some point in time, the capability falls below the required performance level. We can restore the performance before this point, or shortly thereafter. We term such restoration activity as maintenance. There is another situation where we require maintenance. This is when the operator does not know the state of an item, whether it is working or has failed. These are the items that can have hidden or unrevealed failures. In these cases, the role of maintenance is to identify the state by carrying out a test. If the item is in a failed state, we need to carry out further on-failure maintenance to restore it to a working state.

9.1 MAINTENANCE AT THE ACTIVITY LEVEL—AN EXPLANATION OF TERMINOLOGY

9.1.1 Types of maintenance—Terminology and application rationale

When the consequence of failure in service is negligible, we can afford to do the restoration work after the item has failed. We call this strategy on-failure or breakdown maintenance.

Unfortunately, many failures have an unacceptable consequence, so we cannot always apply a breakdown strategy. If we can measure the deterioration and note the period of incipiency, it is possible to predict the time of failure. In such a case, we can schedule the work to ensure minimum disruption of production. This ability to schedule the work facilitates a quick and efficient turnaround. We call this strategy on-condition maintenance, where we can detect and rectify a deteriorating condition before there is functional failure.

In the case of hidden failures, we have to test the equipment periodically. This will identify whether it is in working condition. When we carry out the tests, we carry out failure-finding tasks. If we find the item in a failed state, we rectify it by carrying out breakdown maintenance. Under certain conditions, periodic repair or replacement of the item is warranted, even though it is still in working condition. *Planned maintenance includes all of the following:*

- Testing for hidden failures;
- Condition monitoring of incipient failures;
- Pre-emptive repair or replacement action based on time (running hours, number of starts, number of cycles in operation, or other equivalents of time).

We can summarize the terminology discussed above with the following descriptions of the types of maintenance.

Breakdown Maintenance – repair is done after functional failure of equipment, so it is not possible to schedule the repair work. It is also termed **on-failure maintenance.**

Corrective Maintenance – repair is done after initiation of failure, leading to degraded performance. Usually condition monitoring or inspections will reveal such degradation. The actual repair may be done before or after functional failure, based on our evaluation of consequences of failure, but the key difference from breakdown maintenance is this – we were aware of the functional failure before it occurred, so we had an opportunity to schedule the repair.

Scheduled overhaul or replacement or hard-time maintenance – repair is done based on age (calendar time, number of cycles, number of starts or similar measures of age as appropriate). This strategy is applicable when the age at failure is predictable, i.e., the failure distribution curve is peaky. Fouling, corrosion, fatigue and wear related failures typically exhibit such distributions.

On-condition maintenance – repair is based on the result of inspections or condition-monitoring activities which are themselves scheduled on calendar time to discover if failure has already commenced. Vibration monitoring and on-stream inspections are typical examples of on-condition tasks. Monitoring of some parameters may be continuous, with the use of dedicated instrumentation. All on-condition maintenance is corrective in nature.

Testing or failure-finding is aimed at finding out whether an item is able to work if required to do so on demand. It is applicable to hidden failures and non-repairable items, i.e., the item must be removed from service if we know it has failed. Thereafter, if the item has failed, we do corrective maintenance.

Predictive maintenance – repair is based on predicted time of functional failure, generally by extrapolating from the results of on-condition activities or continuously monitored condition readings. It is synonymous with on-condition maintenance.

Preventive maintenance – repair or inspection task is carried out before functional failure. It is carried out on the basis of age-in-service and the anticipated time of failure. Thus, if the estimate is pessimistic, it may be done even when the equipment is in perfect operating condition. Scheduled overhauls or replacement, on condition and failure finding tasks (themselves time-based), are all part of the preventive maintenance program.

When we do work on a predictive or anticipatory basis, we call it proactive maintenance. If we work on it after it has functionally failed, we call it reactive maintenance. When the incipiency period is relatively small, there is insufficient time available to plan the work. Opportunities to minimize production losses are smaller, and some losses may be unavoidable. In this case, the timing of the work is not in our control, and the corrective maintenance is reactive. Hence corrective maintenance work can be proactive or reactive, depending on the circumstances.

In Chapter 5, we defined planning as the process of thinking through the execution of work. In the course of preparing a plan, we can identify potential pitfalls. We can find solutions in anticipation of the problems, thereby improving the quality and speed of execution. Planned maintenance is that which is correctly prepared sufficiently ahead of its execution. All preventive maintenance can be planned and scheduled.

In most cases, we can plan corrective maintenance as well, but there is less time available to schedule the work, since the onset of failure has already occurred. The term *scheduling* means the allocation of materials and resources as well as assigning a start and finish date to the work.

When it comes to breakdown maintenance however, we do not know the exact scope and timing in advance. It is difficult to plan such work, except in the most generic terms. Hence, breakdown maintenance tends to be less efficient in terms of resource utilization and control of duration.

People tend to regard preventive and predictive maintenance as good while they frown on breakdown maintenance. This view is fashionable but incorrect. It has resulted in unnecessary maintenance expenditure and equipment downtime. There are many failure modes that have little or no effect in terms of consequences on the system or plant as a whole. In such cases, it is economical to allow the failures to take place before taking any action. Preventive maintenance became very popular after the second World War, when the mass production industries enjoyed a period of rapid growth. It became fashionable to apply preventive maintenance strategies as a matter of policy, even in industries where the economic logic was different. The result was that items of equipment became 'due' for maintenance, even though they were performing perfectly well.

There are situations where each of the strategies is appropriate and one must base the selection on the most appropriate way to reduce risks. When the consequences are negligible, the risk is usually low, so a breakdown strategy is appropriate. If there is a threat to safety, production, or the environment, preventive strategies are appropriate.

9.1.2 Applicable maintenance tasks

As the Weibull distribution has wide applicability in maintenance analysis, we will be using the Weibull shape and scale factors in the discussion that follows.

In Chapter 3 (refer Figure 3.16), we discussed the significance of the Weibull shape factor of the ***pdf*** curve. Let us now address the effect of the Weibull shape factor in cases where the failure is evident. When the Weibull shape factor is less than 1, the stresses on the components reduces with time. This can be due to the physical characteristics of the failure mode or to in-built quality problems, and results in an early-failure pattern. When this is a result of underlying quality problems introduced during the design, maintenance, or operational phases, we may do more harm than good by carrying out maintenance. What we need is an analysis of the root cause of the failure, and suitable corrective actions to improve work quality. Similarly when the Weibull shape factor is 1 (or close to 1), the probability of failure does not decrease as a result of planned maintenance work. In this case, we should only do the work when performance has already started deteriorating. We should use the incipiency curve to predict the functional failure. Time-based maintenance strategies are applicable when the Weibull shape factor is >>1, since this indicates a wear out pattern. The higher the value of the Weibull shape factor, the more definite we can be about the time of failure. When this is high, we can easily justify preventive time-based maintenance as it will improve performance. We can determine the maintenance interval by using the ***pdf*** curve to determine the required survival probability at the time of maintenance intervention.

Turning our attention to hidden failures next, we require a time-based test to identify whether the item is in a failed state. If the item has failed already, we have to carry out breakdown maintenance to bring it back in service.

As you can see from the above discussion, only certain tasks are applicable in addressing the failures. The kind of failure, namely, whether it is evident or hidden, and the shape of the ***pdf*** curve help determine the applicable task.

9.1.3 How much preventive maintenance should we do?

The ratio of preventive maintenance work volume to the total is a popular indicator used in monitoring maintenance performance. With a high ratio, we can plan more of the work. As discussed earlier, planning improves performance, so people aim to get a high ratio. In some cases we know that a breakdown maintenance strategy is perfectly applicable and effective. The proportion of such breakdown work will vary from system to system, and plant to plant. There is therefore no ideal ratio of preventive maintenance work to the total. In cases where there is a fair amount of redundancy or buffer storage capacity, we can manage with a very high proportion of breakdown maintenance. In these cases, it will be the lowest total cost option. In a plant assembling automobiles, the stoppage of the production line for a few minutes can prove to be extremely expensive. Here the regime would swing towards a high proportion of preventive maintenance. This is why it is important to analyze the situation before we choose the strategy. The saying, *look before you leap,* is certainly applicable in this context! We have to analyze at the failure mode level and in the applicable operating context. The tasks identified by such analysis would usually consist of some failure modes requiring preventive work, others requiring corrective work, and some others allowed to run to failure. We can work out the correct ratio for each system in a plant, and should align the performance indicators to this ratio.

Figure 9.1 Risk limitation model

9.2 THE RAISON D'ÊTRE OF MAINTENANCE

In Chapter 8, we examined the process of escalation of minor failures into serious incidents. If a serious incident such as an explosion has already taken place, it is important to limit the damage.

We can combine the escalation and damage limitation models and obtain a composite picture of how minor events can eventually lead to serious environmental damage, fatalities, major property damage, or serious loss of production capacity. Figure 9.1 shows this model.

We can now describe the primary role of maintenance as follows:

The raison d'être of maintenance[1] is to minimize the quantified risk of serious safety, environmental, adverse publicity or production incidents that can reduce the viability and profitability of an organization, both in the short and long term, and to do so at the lowest total cost.

This is a positive role of keeping the revenue stream flowing at rated capacity, not merely that of fixing or finding failures. We have to avoid or minimize trips, breakdowns, and predictable failures that affect safety and production. If these do occur, we have to rectify them so as to minimize the severity of safety and production losses. This helps keep the plant safe and profitability high. In the long-term, maintaining the integrity of the plant ensures that safety and environmental incidents are minimal. An organization's good safety and environmental performance keeps the staff morale high and minimizes adverse publicity. It enhances the reputation and helps the organization to retain its right to operate. This assures the viability of the plant. Note that maintenance can reduce the quantified risks, but in the process it can also help reduce the qualitative risks.

Compare this view of maintenance with the conventional view—namely that it is an interruption of normal operations and an unavoidable cost burden. We recognize that every organization is susceptible to serious incidents that may result in large losses. Only a few of the minor events will escalate into serious incidents, so it is not possible to predict precisely when they will occur. One could take the view that one cannot anticipate such incidents. Often, we can see that the situation is ripe and ready for a serious incident, as in the case of Piper Alpha, but even so, we cannot predict the timing.

Sometimes these losses are so large that they may result in the closure or bankruptcy of the organization itself As an example from a service industry, consider the collapse of the Barings Bank[2]. Their Singapore branch trader Nick Leeson speculated heavily in arbitraging deals, losing very large sums of money in the process. He did this over a relatively long period of time, using a large number of ordinary or routine looking transactions. There were deviations from the Bank's policies, which an observant management could have noticed. In our model, these deviations from the norm constitute the process demand rate. Leeson was a high performing trader, and in order to operate effectively, he needed to make quick decisions. So the Bank removed some of the normal checks and balances. These controls included, for example, the separation of the authority to buy or sell on the one hand, and on the other hand, to settle the payments.Thus, they defeated a *Procedure* barrier, permitting an opportunity for event escalation. With the benefit of hindsight, we can question whether the reliability of the *People* barrier was sufficiently high to justify this confidence. Barings had carried out an internal audit a few months before

Leeson's activities came to light. In terms of our model, this was a test to identify hidden failures. The auditors did find some areas of concern, and recommended that Leeson's authority be limited to trading or settlements, but not both. The Bank did not implement this recommendation. By January 1995 the London office was providing more than $10 million per day to cover the margin payment to the Singapore Exchange. There were clear indications that something was amiss, but all the people involved ignored them. The Bank of England, which supervised the operations of Barings Bank, wondered how Barings Singapore was so profitable but did not pursue the matter further. Hence the *People* barrier in the damage limitation level was also weak. When you compare this disaster with Piper Alpha, Bhopal, or Chernobyl, some of the similarities become evident. With so many barriers defeated, a disaster was looming, and it was only a matter of time before it happened.

Integrity issues are quite often the result of unrevealed failures. We can minimize escalation of minor events by taking the following steps:

- Reduce the process variability to reduce the demand rate;
- Increase the barrier availability. We can do this by increasing the intrinsic reliability, through an improvement in the design or configuration. Alternatively, we can increase the test frequency to achieve the same results;
- Do the above in a cost effective way.

We discussed the effect of the law of diminishing returns and how to determine the most cost-effective strategy, in Chapter 8. In order to achieve the required level of availability in the case of each barrier, we have to determine its intrinsic reliability. We can then calculate the test interval to produce the required level of availability.

At this stage, we encounter a practical problem. How does one measure the reliability of the *People or Procedures* barriers? There is no simple metric to use, and even if there was one, a consistent and repeatable methodology is not available. If we take the case of the *People* barrier, their knowledge, competence, and motivation are all important factors contributing to the barrier availability. As we discussed in Chapter 8, motivation can change with time, and is easily influenced by unrelated outside factors.

There would be an element of similarity in motivation due to the company culture, working conditions, and the level of involvement and participation. As long as the average value is high and the deviations small, there is no problem. Also, if there are at least two people available to do a job in an emergency, the redundancy can help improve the barrier availability. We can test the knowledge and competence of an individual from time to time, either by formal tests or by observing their performance under conditions of stress. In an environment where people help one another, the *People* barrier availability can be quite high. In this context, salary and reward structures that favor individual performance in contrast to that of the team can be counterproductive.

Procedures used on a day-to-day basis will receive comments frequently. These will initiate revisions, so they will be up-to-date. Those used infrequently will gather dust and become out-of-date. If they affect critical functions, they need more frequent review. We should verify *Procedures* relating to damage limitation periodically, with tests (such as building evacuation drills).

The predominance of soft issues in the case of the *People* and *Procedures* barriers means that estimating their reliability is a question of judgment. Redundancy helps, at least up to a point, in the case of the *People* barrier. Illustrations, floor plans, and memory-jogger cards are useful aids in improving the availability of the *Procedures* barrier. It is a good practice to keep some drawings and procedures permanently at the work site. Thus we see some wiring diagrams on the doors of control cabinets. Similarly we get help screens with the click of a mouse button and see fire-escape instructions on the doors of hotel rooms. Obviously, we have to ensure that these are kept up to date by periodic replacement.

9.3 THE CONTINUOUS IMPROVEMENT CYCLE

Once the plant enters its operational phase, we can monitor its performance. This enables us to improve the effectiveness of maintenance. This process can be represented by a model, based on the Shewhart[3] cycle.

In this model, we represent the maintenance process in four phases. The first of these is the planning phase, where we think through the execution of the work. In this phase, we evaluate alternative maintenance strategies in terms of the probability of success as well as costs and benefits. In the next phase, we schedule the work. At this point, we allocate resources and finalize the timing. In the third phase, we execute the work, and at the same time we generate data. Some of this data is very useful in the next phase, namely that of analysis, and we will discuss the data we need and how to collect it in Chapter 11. The results of the analysis are useful in improving the planning of future work. This completes the continuous improvement cycle. Figure 9.2 below shows these four phases.

9.3.1 Planning

We begin the planning process by defining the objectives. The production plant has to achieve a level of system effectiveness that is compatible with the production targets. We have to demonstrate that the availability of the safety systems installed in the plant meets the required barrier availability. Using reliability block diagrams, we can translate these requirements to availability requirements at the sub-system and equipment level.

The next step consists of identifying those failure modes that will prevent us from achieving the target availability. Next, we evaluate alternative ways to

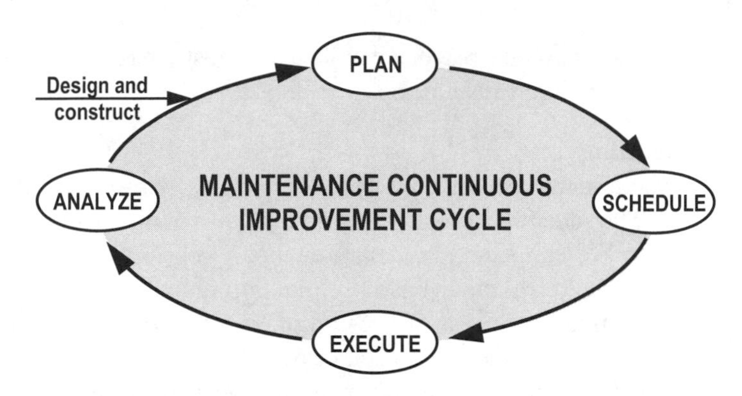

Figure 9.2 Continuous improvement cycle.

resolve these problems. We have to execute the selected tasks at the correct frequencies, with the specified skilled resources. We can bundle a number of these tasks together. We can do so if the work is on the same equipment, using the same trade skills at the same frequency. We call such an assembly of tasks a maintenance routine. These routines will cover all time-based tasks including condition monitoring and failure finding tasks.

When we execute condition monitoring tasks, we will detect incipient failures. This will result in the generation of corrective maintenance work. We carry out failure finding tasks, to identify whether items subject to hidden failures are in a working state. If they are in a failed state, we have to carry out breakdown maintenance work to restore it to a working condition. Lastly, we will allow certain items of equipment to run to failure, and some others will fail in service as a result of poor operation or maintenance. These will also require breakdown maintenance. We have to make a provision for such corrective and breakdown work in our plan. Various tools are available to assist us in planning this work, and we will review some of these in the next chapter.

We cannot execute all the work during normal operations and so some of these will require a plant shutdown.

Planning of maintenance encompasses all the routine and corrective work done during normal operations as well as during shutdowns. There is an element of generic planning that we can do with respect to breakdowns. For example, in a plant using process steam, we can expect leaks from flanges, screwed connections, and valve glands from time to time. These leaks can grow rapidly, especially if the pressures involved are high, or the steam is wet. The prompt availability of leak-sealing equipment and skilled personnel can prevent the event from escalating into a plant shutdown. In the case of plans made to cope with breakdowns, the work scope is usually not definable in advance. We require a generic plan that will cater to a variety of situations. Note that while such a plan may be in place, we still cannot schedule the work till there is a failure. If a breakdown does take place, we will have to postpone some low priority work, so that we can divert resources to the breakdown.

9.3.2 Scheduling

We have to schedule maintenance work in such a way that we minimize production losses. The scheduler's task is to find windows of opportunity to minimize the losses. We can schedule maintenance work during weekends or month-ends if there are calendar-based production quotas. We schedule the work so that it commences towards the end of the week or month, and complete it in the early part of the next week or month. By boosting the production rate before and after the transition point, we can build up sufficient additional production volumes to compensate for the production lost during the maintenance activity.

We can avoid loss of production if intermediate storage or installed spares are available. When carrying out long duration maintenance work on protective system equipment such as fire pumps, the scheduler must evaluate the risks and take suitable action. For example, we can bring in additional portable equipment to fulfill the function of the equipment under maintenance. If this is not possible, we have to reduce the demand rate, for example, by not permitting hot work. Using this logic, one can see why the Piper Alpha situation was vulnerable. The fire deluge systems were in poor shape, the fire pumps on manual, at a time when there was a high maintenance and project workload with a large volume of hot work.

We have to prioritize the work, with jobs affecting integrity at the highest level. This means that testing protective devices and systems has the highest priority. Work affecting production is next in importance. Within this set, we can prioritize the work according to the potential or actual losses. All other work falls in the third category of priorities. When scheduling maintenance work, we have to allocate resources to the high priority work and thereafter to the remaining work. If the available resources are inadequate to liquidate all the work on an ongoing basis, we have to mobilize additional resources. We can use contractors to execute such work as a peak-shaving exercise.

The available pool of skills may not meet the requirements on a day-to-day basis. If each person has a primary skill and one or two other skills, scheduling becomes easier. This requires flexible work-practices and a properly trained workforce. On the other hand, if restrictive work practices apply, scheduling becomes more difficult.

We then have to firm up the duration and timing of each item of work, arrange materials and spare parts, special tools if required, cranes and lifting gear, and transportation for the crew. When overhauling complex machinery, we may need the vendor's engineer. Similarly we may require specialist machining facilities. We have to plan all these requirements in advance. It is the scheduler's job to ensure that the required facilities are available at the right time and place and to communicate the information to the relevant people.

A good computerized maintenance management system (CMMS) can help us greatly in scheduling the work efficiently.

9.3.3 Execution

The most important aspects in the execution of maintenance work are safety and quality. We have to make every effort to ensure the safety of the workers. Toolbox talks, which we discussed in Chapter 6, are a good way of ensuring two-way communications. They are like safety refresher training courses. A more formal Job Safety Analysis (JSA), used in some high hazard industries helps increase safety awareness in maintenance and operational staff. JSA cards are used not just for hazardous activities, they are also used for increasing awareness during routine maintenance activities. The worker needs protective apparel such as a hard hat, gloves, goggles, overalls, and special shoes. These ensure that even if an accident occurs, there is no injury to the worker. Note that protective apparel is the *Plant* barrier in this case. If the work is hazardous, for example, involving the potential release of toxic gases, we must ensure that the workers use respiratory protection. In cases where the consequence of accidents can be very high, escape routes needs advance planning. We have noted earlier that redundancy increases the availability of *Plant*. Hence in high risk cases, we should prepare two independent escape routes. In addition to the normal toolbox talk, the workers should carry out a dry run before starting the hazardous work. During this dry run, they will practice their escape in full protective gear. The damage limitation barriers must also be in place. For example, in the case discussed above, we must arrange standby medical attention and rescue equipment. In a practical sense, the management of risk requires us to ensure that the *People, Plant,* and *Procedure* barriers are in-place and in good working condition.

The quality of work determines the operational reliability of the equipment. In order to reach the intrinsic or built-in reliability levels, we must operate the equipment as designed, and maintain them properly. Both require knowledge, skills, and motivation. One can acquire knowledge and skills by suitable training. We can test and confirm the worker's competence. Pride of ownership and motivation are more difficult issues, and they require a lot of effort and attention. The employees and contractors must share the values of the organization, feel that they get a fair treatment, and enjoy the work they are doing. This is an area in which managers are not always very comfortable. As a result, their effort goes into the areas in which they are comfortable and they tend to concentrate on items relating to technology, knowledge and skills. Quality is a frame of mind, and motivation is an important contributor.

Good planning and organization are necessary for efficient execution of work. A number of things must be in-place, in good time. These include the following:

- Permits to work;
- Drawings and documentation;
- Tools;
- Logistic support, spare parts, and consumables;
- Safety gear;
- Scaffolding and other site preparation.

If these are not in place, we will waste resources while waiting for the required item or service. The efficiency of execution is dependent on the quality of planning and organization.

The two drivers of maintenance cost are the operational reliability of the equipment, and the efficiency with which we execute the work. We require good quality work from both operators and maintainers to achieve high levels of reliability. The number of maintenance interventions falls as the reliability improves. This also means that equipment will be in operation for longer periods. When we carry out maintenance work efficiently, there is minimum wastage of resources. As a result, we can minimize the maintenance cost. As we have already noted, good work quality improves equipment reliability, and good planning helps raise the efficiency of execution. These two factors, work quality and good planning, are where we must focus our attention.

There are many reasons for delays in commencing the planned maintenance work. There may be a delay in the release of equipment due to production pressures. Similarly, if critical spares, logistic support, or skilled resources are not available, we may have to postpone the work to a more convenient time. While we can tolerate some slippage, it is counter productive to spend a lot of time and money deciding when to do maintenance, and then not do it at the correct time. When planned work is done on schedule, we have achieved compliance. For practical purposes, we accept it as compliant as long as it is completed within a small range, usually defined as a percentage of the scheduled interval. As a guideline, we should commence items of work that we consider safety critical, within +/-10% of the planned maintenance interval, from the scheduled date. For safety critical work that is planned every month, e.g., lubricating oil top-up of the gear-box of fire pumps, we would consider it compliant if it was executed some time between 27 and 33 days from the scheduled date on the previous occasion. If the work was considered production critical, again planned as a monthly routine, e.g, lubricating oil top-up of the gear-box of a single process pump, as long as the work was done within +/-25%, or in this case between 23 and 37 days of the previous due date, it would be considered compliant. Finally, if the same work was planned on non-critical equipment, e.g., the gearbox of a duty pump (with a 100% standby pump available), a wider band of, say +/-50% is acceptable. In this case, for a monthly routine, if the work was done between 15 and 45 days of the previous scheduled date, it would be considered compliant. Progressive slippage is not a good idea. Thus, we must retain the original scheduled dates even if there was a delay on the previous occasion. If the work falls outside these ranges, the maintenance manager must approve and record the deviations. This step will ensure that we have an audit trail.

Procedural delays, caused for example, by having a permit-to-work system that needs a dozen or more signatures are sometimes encountered. The Author has audited one location where technicians sat around every morning for 1.5-2 hours, waiting for the permits-to-work. No work started before this time, and the site considered this practice normal. The PTW for simple low-hazard activities needed 12 signatures, mostly to 'inform' various operating staff that work was going on. Over the years, the PTW had evolved into a work slowdown process, instead of being the enabler of safe and productive work.

The timely execution of work is very important, so we should measure and report compliance. This is simply a ratio of the number of jobs completed on the due date (within the tolerance bands discussed earlier), to those scheduled in a month, quarter, or year. This is a key performance indicator to judge the output of maintenance.

We noted earlier that whenever we do work, we generate data. Such data can be very useful in monitoring the quality and efficiency of execution. By analyzing this data, we can improve the planning of maintenance work in future, as discussed below.

9.3.4 Analysis

The purpose of analysis is to evaluate the performance of each phase of maintenance work—planning, scheduling, and execution. The quality and efficiency of the work depend on how well we carry out each phase. There is a tendency to concentrate on execution, but if we do not look at how well we plan and schedule the work, we may end up doing unnecessary or incorrect work efficiently!

In the planning phase, it is important to ensure that we do work on those systems, sub-systems, and equipment that matter. Failure of these items will result in safety, environmental, and production consequences. How well we increase the revenue streams and decrease the cost streams determines the value added. Quite often, the existing maintenance plan may simply be a collection of tasks recommended by the vendors, or a set of routines established by custom and practice. So we may end up doing maintenance on items whose failures do not matter.

The objective of planning is to maximize the value added. We do this by carrying out a structured analysis to establish the strategy at the failure mode level. This task can be large and time-consuming, so we have to break it up into small manageable portions. We must analyze only those systems that matter, therefore that we use our planning resources effectively. We identify progress milestones after estimating the selection and analysis workload. In effect, we make a plan for the plan. To achieve this objective, we have to measure the progress using these milestones. Such an analysis can help monitor the planning process.

At the time of execution, we may find that some spare part, tool, resource, or other requirement is not available. This can happen if the planner did not identify it in the first place or the scheduler did not make suitable arrangements. There will then be an avoidable delay. We can attribute such delays to defective planning or scheduling. A measure of the quality of planning and scheduling is the ratio of the time lost to the total.

In the execution phase, we can identify a number of performance parameters to monitor. The danger is that we pick too many of them. In keeping with our objectives, safety and the environment are at the top of our list, therefore we will measure the number of high potential safety and environmental incidents. We discussed the importance of hidden failures in the context of barrier availability. We maintain system availability at the required level by testing those items of equipment that perform a protective function. Operators or maintainers may carry out such tests, the practice varying from plant to plant. The result of the test is what is important, not who does it. We have to record failures as well as successful tests. Sometimes people carry out pre-tests in advance of the official tests. Pre-tests defeat the objective of the test, since the first test is the only way to know if the protective device would have functioned in a real emergency. In such a case, we should report the results of the pre-test as if it is the real test, so that the availability calculations are meaningful. If a spurious trip takes place, this is a fail-to-safe event. By recording such spurious events, we can carry out meaningful analysis of these events.

One can use some simple indicators to measure the quality of maintenance. These include, for example, the number of days since the last trip of the production system, sub-system, or critical equipment. Another measure is the number of days that critical safety or production systems are down for maintenance. If we concentrate on trends, we can get a reasonable picture of the maintenance quality. Note that work force productivity and costs do not feature here, as safety and quality are the first order of business.

Earlier, we discussed the importance of doing the planned work at or close to the original scheduled time. Compliance is an important parameter that we should measure and analyze. The ratio of planned work to the total, and associated costs are other useful indicators. In measuring parameters such as costs, it is useful to try to normalize them in a way that is meaningful and reasonable, to enable comparison with similar items elsewhere. For this purpose, we use some unit representing the complexity and size of the plant such as the volumes processed or plant replacement value in the denominator.

Finally, we can evaluate the analysis phase itself, by measuring the improvements made to the plan as a result of the analysis. In a Thermal Cracker unit in a petroleum refinery, the six-monthly clean-out shutdowns used to take 21 days. Over a period of three years, the shutdown manager reduced the duration to 9 days, while stretching the shutdown intervals to 8 months. The value added by this plant was $60,000 per day, so these changes meant that the profitability increased by about $1.7 million per annum. This required careful analysis of the activities, new ways of working, and minor modifications to the design to reduce the duration and increase the run lengths. The plant was located in the Middle East, where day temperatures could be 40 - 50°C. Working inside columns and vessels under these conditions could be very tiring and, therefore, took a long time. One suggestion was to cool the fractionator column and soaker vessel internally, using a portable air-conditioning unit. In the past, they had been used to cool reactors in Hydro-Cracker shutdowns, to reduce the cooldown time. Use of these units for the comfort of people was a new application. When the shutdown manager introduced air-conditioning, the productivity rose sharply, and this helped reduce the duration by about 36 hours. Another change was to relocate two pairs of 10 inch flanges on transfer lines from the furnace to the soaker. This clipped an additional six hours. There were many more such innovations, each contributing just a few hours, but the overall improvement was quite dramatic. This case study illustrates how one can measure the success of the analysis phase in improving the plan and thus the profitability.

It is easy to fall into the trap of carrying out analysis for its own sake. In order to keep the focus on the improvements to the plan, we need to record changes to the plan as a result of the analysis. Further, we have to estimate the value added by these changes and bank them. Hence, analysis must focus on improvements to all four phases of the maintenance process.

9.4 SYSTEM EFFECTIVENESS AND MAINTENANCE

The primary role of maintenance is to minimize the risk of minor events escalating into major incidents. We achieve this by ensuring the required level of barrier availability. Let us examine how we can do this in practice, with some examples.

9.4.1 Testing of pressure relief valves

Pressure relief valves (PRVs) are important protective devices. They protect the vessel or piping from over pressure and potential disaster. In most cases, there is no redundancy built in, and each PRV must perform when there is a demand. Normally, there are no isolation valves on the inlet and outlet of single PRVs. In such cases, unless we find a way to test them in service, the only opportunity is when we decommission the associated vessel or pipeline. The flip side is that if we have to test the PRV, we have to take the vessel out of service. In most cases, we cannot decommission vessels without a plant shutdown. This means that the test frequency of the limiting PRVs often determines the periodicity of the shutdowns. This goes against the attempts to increase the intervals between shutdowns.

In the case of hidden failures, it is not easy to determine the exact failure distribution of a single item. Therefore, we make some simplifying assumptions, as follows:

- The failure distribution is exponential;
- Similar items in broadly similar service fail in the same manner.

Under these conditions, the hazard rate is constant and we call it the failure rate. It is unlikely that there will be a sufficient number of failures on a single PRV to be able to calculate its failure rate. The common practice is to collect failure data for a family of PRVs of a given type, in a given service. For example, we could collect failure data for balanced-bellows PRVs in hydrocarbon gas service. If the population of PRVs is large, we can sort the data set by type of fluid, pressure range, and by make and model. However, as we try to narrow down and refine the data set, the sample size becomes smaller, reducing the confidence level in the calculated failure rate. Note that the failure rate we are considering here is the fail-to-danger rate, or the failure to lift at 110% of the cold set pressure. For a given sample of PRVs tested on the bench, we count the number of PRVs that do not lift when the test pressure is 110% or more of the cold set pressure. The cumulative operating period is the sum of the periods that each of the PRVs in the sample has been in service. Dividing the number of failures by the cumulative operating period gives the failure rate.

In some plants, the designer may have provided two PRVs each with 100% relieving capacity, in a one-out-of-two configuration. In this case, there are two PRV positions, with inter-locked isolation valves. If the test interval is limiting the shutdown intervals, one solution is to install both PRVs and leave their inlet and outlet isolation valves permanently open. It is advantageous to stagger the cold set pressures of the two PRVs slightly, typically by 1-2%. This will ensure that one PRV will always lift first, and the second one will only come into operation if the first one fails to lift. Figures 9.3 and 9.5 illustrate the two alternative designs, along with their RBDs in Figures 9.4 and 9.6 respectively.

If the failure rate of the PRV is 0.005 per year (or an MTTF of 200 years), and the required mean availability is 99.5%, the test interval in the single PRV case is 2.01 years, using expression 3.13. In the second case, the system as a whole, with two PRVs in parallel should now have a mean availability of 99.5%. This case is similar to that in expression 8.2, but with two parallel blocks in this RBD. The required availability can be calculated thus,

$$(1 - A_{system}) = (1 - A_{prv1})(1 - A_{prv2}) \qquad 9.1$$

The two blocks are identical, so $A_{prv1} = A_{prv2}$. What will be the availability of the protective system as a whole with this configuration? Table 9.1

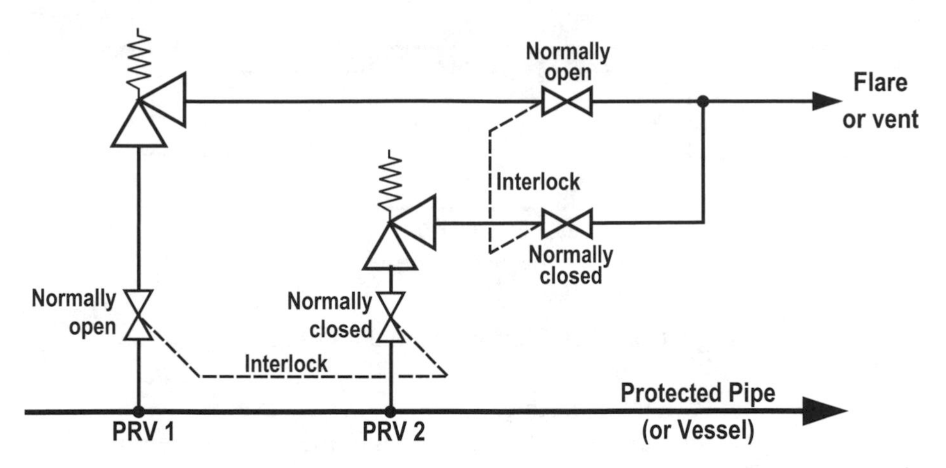

Figure 9.3 Conventional arrangement of spared PRVs.

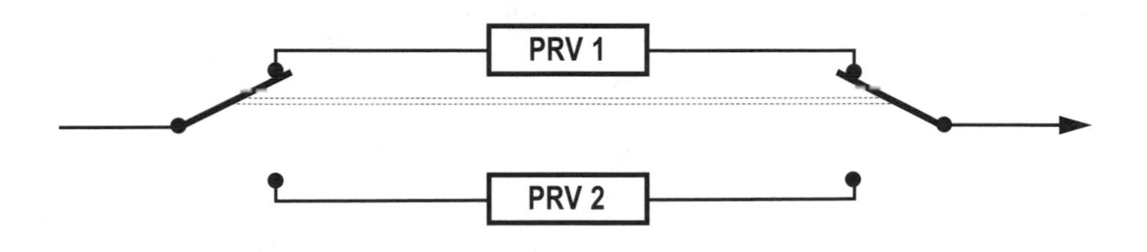

Figure 9.4 RBD for arrangement in Fig. 9.3.

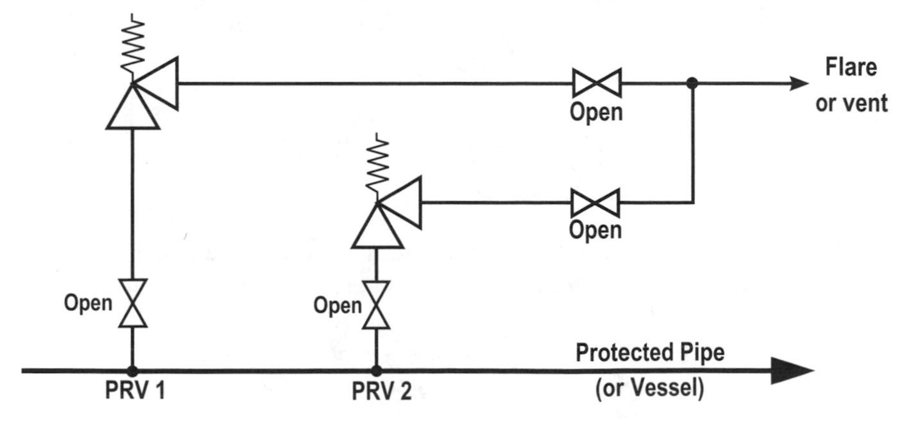

Figure 9.5 Alternative arrangement of spared PRVs.

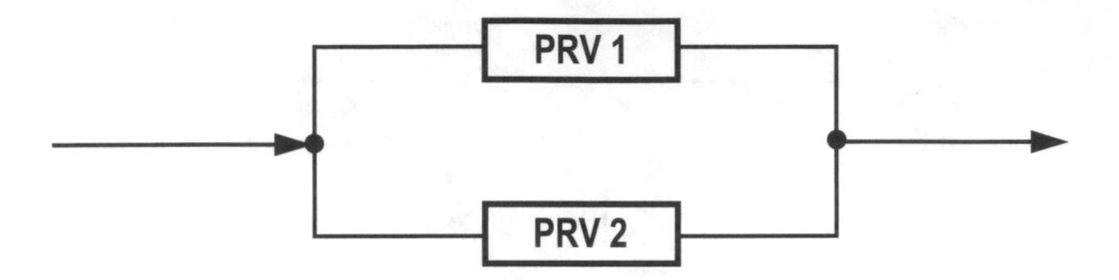

Figure 9.6 RBD for arrangement in Fig. 9.5.

shows the effect of different test intervals on the system availability. Since the desired system availability is 0.995, we could, in theory, manage this with a test interval of 30 years. In practice, we would select a test interval of 3 or 4 years, and check the effect on failure rates. This example demonstrates that the effect of redundancy is quite dramatic. Merely by making a change in operational philosophy, both PRVs can, in theory, move to a significantly larger test interval. As a result of the systems approach, the PRVs need no longer be the limiting case when determining shutdown intervals.

MTTF years	Test Intvl. years	T/MTTF	Exp - T/MTTF	Availability of		
				PRV 1	PRV 2	SYSTEM
200	1	0.0050	0.995012	0.997506	0.997506	0.999994
200	1.5	0.0075	0.992528	0.996264	0.996264	0.999986
200	2	0.0100	0.99005	**0.995025**	**0.995025**	0.999975
200	2.5	0.0125	0.987578	0.993789	0.993789	0.999961
200	3	0.0150	0.985112	0.992556	0.992556	0.999945
200	3.5	0.0175	0.982652	0.991326	0.991326	0.999925
200	4	0.0200	0.980199	0.990099	0.990099	0.999902
200	5	0.0250	0.97531	0.987655	0.987655	0.999848
200	7	0.0350	0.965605	0.982803	0.982803	0.999704
200	10	0.0500	0.951229	0.975615	0.975615	0.999405
200	15	0.0750	0.927743	0.963872	0.963872	0.998695
200	20	0.1000	0.904837	0.952419	0.952419	0.997736
200	30	0.1500	0.860708	0.930354	0.930354	**0.995149**

Notes:

Both PRVs assumed to perform similarly and follow exponential distribution.
Both PRVs have the same MTTF for failing to lift at 110% of set pressure, of 200 years.
With one PRV in service, with a 2-year test interval, the PRV availability is 0.995.
With one PRV in service, with a 2-year test interval, the System availability is 0.995.
With both PRVs in service, at a test interval of 30 years, the System availability is 0.995.

Table 9.1 Mean availabilty with alternative configuration of spared PVRs.

If there are multiple-barrier protection systems, we can take credit for them in the same manner as in the case of the system discussed above. A plant may have separate blow-down, emergency shutdown, and pressure control systems. These, along with the PRVs, provide pressure protection. As long as each system is independent, we can represent the systems as parallel blocks in the RBD.

A word of caution is in order at this point. The process pressure actuates pressure control systems and PRVs. A rise in operating pressure may also initiate actions on other protective systems, such as emergency shutdown or fire protection systems. Initiating signals from e.g., the fire detection system, can trigger these systems, but may not trigger the lifting of PRVs. The location of the pressure sensing element may be remote from the vessel being protected. In the case of pool or jet fires that affect the vessel in question, the other devices may not respond as quickly as the PRVs. Hence, when we seek credit for multiple barrier systems, we must consider each type of incident on merit. A second word of caution is also in order. Among the reasons for PRVs failing is fouling, caused by the process fluid. Since fouling rates of PRVs are not very predictable, we cannot increase test intervals indefinitely and one must use good judgment before making any changes.

The discussion so far has been with respect to protection against over-pressure. Spurious operation of the PRV is also unacceptable when the process fluid is toxic or flammable, or could damage the environment. We can calculate the PRV test frequency for this scenario as well, using expression 3.13. In this case, the process demand rate will depend on how close the operating pressure is to the cold set pressure, and the steadiness or otherwise of the process. The required PRV availability is dependent on this demand rate.

We call these fail-to-safe events because over-pressure cannot take place. In the special circumstances discussed above, leakage of process fluids may be harmful, so the terminology is unfortunate. We use this failure-rate in the calculation and obtain it as follows:

$$\frac{\textit{Number of PRVs that lift or leak below 90\% cold set pressure}}{\textit{Cumulative operational period of all the PRVs in the sample}} \qquad 9.2$$

Note that the PRV is tested on the bench before overhaul, and again after the cleaning, repair, and resetting. The failure rates of interest are those obtained in the pre-overhaul tests. The results tell us what could have happened in the plant had the PRVs remained in service.

The actual failure rate of the PRV in the installed location can be different from that measured on the test bench, for a number or reasons, including the following:

- Forces, torsion, or bending moments on the PRV body, as a result of pipe stresses at site;
- Mechanical damage caused to the PRV in transporting it to the test bench;

- Displacement of scale or gumming material during transport to the test bench.

It is good engineering practice to measure the displacement of the PRV discharge pipe flange, when we open that joint. The pipe flange may move away from the PRV flange axially or transversely. It may wedge open, and the flange face gap becomes larger on one side than on the opposite side. There can also be rotational misalignment of the flange bolt holes due to fabrication errors. Some combination of all three types of misalignment is possible. The result of such defects will be to cause a force, moment, or torque on the PRV body. PRVs are delicate instruments, and their settings can change as a result of these stresses. When this happens, we can expect the PRV to leak or lift before reaching the cold set pressure, resulting in spurious operation.

PRVs need care in handling, especially during transportation. When moving them to and from the work site, it is a good practice to bolt them firmly on a pallet or transport housing, with the inlet and outlet capped off with plastic bungs. When we remove a PRV from its location for testing, it is not possible to guarantee that scale or deposits in the inlet or outlet nozzles remain undisturbed during transportation. As long as we handle the PRV with care, we can minimize the displacement of deposits. If possible, we should try to minimize the handling by doing the pre-overhaul tests close to the work site.

9.4.2 Duty-standby operation

The purpose of standby equipment is to ensure a high level of process system availability. The configuration may be 1 out of 2 (1oo2), 2oo3, 3oo4 or similar. A common operating practice is to run standby equipment alternately with the duty equipment, so that in most cases the running hours are roughly equal. This practice has some benefits from the operational point of view, as listed below.

- The operators know that both the duty and standby equipment work, because they have witnessed both in running condition;
- The equipment accumulates equal running hours, and operating experience;
- In some cases, start-up procedures are difficult and time consuming. Once the standby starts up, it is convenient to leave it running, and not have to restart the original equipment.

In the days before the introduction of mechanical seals, packed glands provided shaft sealing in reciprocating and rotating machinery.The packing needed regular lubrication or it would dry up and harden, making it useless. In the majority of cases, the only way to lubricate the packing was to run the equipment, allowing the process fluid to provide the lubrication. The practice of running duty and standby equipment alternately met this requirement. The practice still continues, even though mechanical seals have largely replaced the packed glands long ago.

Mechanical seal failures form a significant proportion of the total. The wear of the seal faces takes place mainly during their start-up phase. At this time, the hydrodynamic fluid film is not yet in place, and the seal runs dry. After a short while, the fluid film is established, separating the seal faces and reducing wear. Frequent starts are a major cause of wear in seals, and by reducing the number of starts, we can reduce the number of seal failures, and hence pump failures.

Let us consider the case of a 1oo2 pumping situation, where we have a designated duty and standby pump.The consequences of failure of the two pumps differ, as the following argument shows. If the duty pump fails in service, the standby cuts in and in most cases there is no impact on production. On the basis of the production consequence of failure, it is difficult to justify any maintenance work on the duty pump. If the direct maintenance cost of failure is high, we can justify a limited amount of preventive maintenance, typically condition monitoring.

If the standby pump does not start on demand, it has serious consequences. Its only role is to start if the duty pump fails, and take over the full pumping load.

This is a hidden failure, and the remedy is to test start the standby pump. At what frequency shall we carry out the test? Depending on historical failure rates relating to this failure mode, we can test start it at a suitable frequency (using expression 3.13), to obtain the desired availability.

The next functional failure to consider is the inability to deliver the required flow at the operating pressure. To check this condition, we test the standby equipment on full load for 4 to 24 hours. A spin-off benefit from running long duration full load tests is that it will then be possible to take condition monitoring readings for the standby equipment regularly.

Now consider the situation when we run the pumps alternately - either pump, if running at the time, may fail while running. If on standby, it may not start or perform satisfactorily. Thus both pumps need maintenance, often with poor condition monitoring data (since the collection of data is a hit or miss affair). The wear out rate is about equal, and the conservative policy would be to carry out condition- or time-based overhauls on both pumps. This is costly and inefficient. Last, with a similar level of wear-out taking place on each pump, they are both equally likely to fail, and thus will become worse with time. The advantage of having a redundant system is therefore greatly reduced.

The operating policy of alternate duty operation results in many starts, which tends to increase seal failure rates. This in turn means that there is an increase in the level of risk. In the case of duty/standby operation, the test frequencies will generally be quite low, and hence we require fewer total starts. The failure rates will therefore be lower, and it is the option with the lower risk.

There are two outcomes to consider, one relating to up-time and the second relating to costs. In the case where we run both pumps alternately, we have to take both out of service from time to time, to carry out overhauls. In the duty-standby case, only when the duty pump exhibits performance problems do we initiate maintenance work. Similarly, we will work on the standby pump only if the test run fails. We can see that the total downtime will be higher in the case where the duty is alternating. Due to the higher seal failure rate, in absolute terms the workload will be higher. Further, the longer the downtime on one pump, the greater the chances that the other will fail while running. Overall, the system availability will tend to fall.

In systems with installed spares, the availability will be higher when we designate duty and standby equipment, and align the operating policy suitably. The reduced maintenance workload has an immediate favorable impact on maintenance costs. We have seen cost and uptime improvements of 10% or more merely by switching to a duty-standby philosophy.

In some cases, the equipment start-ups are quite difficult. Once started up, it is often prudent to leave the equipment running. In these cases, we cannot follow a strict duty-standby regime. The solution is to operate the duty and standby equipment unequally on a 90:10 or 75:25 basis. In this case, we run the duty equipment for, say, three months, and the standby equipment for, say, one month. The advantage of this policy is that it produces a low number of starts, while allowing a long duration test run (of one month), and a long test interval (of three months). We can determine the actual frequency in each case using expression 3.13. We can round up (or down) the test frequencies for administrative convenience.

Equipment such as gas turbines have dominant failure modes that are reasonably predictable. The vendors provide charts for derating the interval between major overhauls. The derating factors depend on the number of starts and loads on the machines. Gas turbine drivers of electrical power generators are invariably in systems with built-in redundancy. We have to work out the timing of their overhauls very carefully. One of the determining factors is the availability of (the high cost) spare parts. The vendor reconditions these spare parts off-line, so there is a known lead-time involved in obtaining them. It is therefore advantageous to plan their overhauls with this constraint in mind. We plan the operation of gas turbines to suit the reconditioning cycle of critical spares.

9.4.3 End-to-end testing of control loops

A control loop has three main elements: the sensing device, the control unit, and the executive device, as shown in Figure 9.7 below. For the purpose of this discussion, we include the cable or tubing termination in the relevant element, and ignore failures of the cable or tubing. Sensing devices measure flow, pressure, temperature, speed, smoke density, and vibration levels. The control unit or black box compares the inputs received from the sensing device with a control setting or logic. It then produces an output signal designed to bring the process back into control, or shut down the system safely. The complexity of control of units can vary, with software being used extensively in modern units. The executive device can be similar to the following types of devices:

- A simple control valve;
- An electrically or hydraulically operated emergency shutdown valve;
- A trip and throttle valve on a steam turbine;
- The hydraulic actuator of the rudder of a ship;
- The trip-actuated valve in a deluge system.

Figure 9.7 RBD of control system.

When dealing with hidden failures, it is necessary to test the relevant control loops. Safety systems are often subject to hidden failures. There may be significant production losses and additional maintenance work as a result of these tests. However, it is easy to test parts of the system at low cost, so we often adopt this method. Sensing and control units are susceptible to drift and span changes, which will result in incorrect output signals. We can test the sensing units by defeating the outputs from the control unit for the duration of the test. We can thus establish the availability of the sensing units. We can supply a variety of input signals to the control units, and measure the outputs. As before, we disconnect the executive unit from the control unit, so that we can avoid executive action. The test demonstrates that the control unit generates the required executive signals, thereby establishing its availability. Finally, we come to the action of the executive unit itself. This is the final element in the chain. The production losses referred to earlier relate to the action of this final element. The closure of an emergency shutdown valve results in production losses. We have to avoid or minimize the losses, without forgoing the test. One way of doing this is to permit partial rather than complete closure.We can be do this by providing, for example, a mechanical stop that limits the travel of the valve. Apart from the fact that we minimize production losses, such partial closure tests reduce the wear and tear on the valve. This is especially important in the case of valves with soft seals. Their function is to stop the process flow during a real emergency, so we cannot afford to damage them by inappropriate tests. The failure rates depend on the number of operations of the valve. When a valve has to open from a fully closed position, at the time of opening it has to open under the full differential pressure. It requires large forces or torques to crack open the valve. Thereafter, the differential pressure falls, and the loading reduces. Hence, a total closure can cause significant damage to the seats, while a partial one does not do as much damage. The fact that the valve moves, even by a small amount, is enough to prove to us that it is in working condition. There is a small chance with partial closure tests that the valves may not close fully, when called upon to do so. Therefore, we have to back up such partial closure tests by a less frequent total closure test. We can do this whenever possible, for example, just prior to a planned shutdown of the plant. When selecting valves, their ability to survive full closure tests should be given due importance.

The control units are susceptible to another kind of failure, attributable to poor change control procedures. With the increasing use of software, we can alter the logic fairly easily. We have only to modify the lines of code affecting, for example, the set points. *There is a distinct possibility of loss of control, so we must insist on rigorously using the change control procedure for such changes.* Trained and competent people must carry out these alterations. One must verify the quality of the change with a suitable verification routine. The normal test regime used for demonstrating the availability of the control loop is not the means for doing such verification. In section 5.6, we discussed the Flixborough disaster, where they carried out piping changes without using a rigorous change control procedure. Unauthorized software changes could cause another Flixborough. *High performance organizations enforce them rigorously, and carry out (external) audits periodically to capture any deviations. Software changes are inherently difficult to locate, so additional control steps are required. When people understand why change control matters, we can prevent unauthorized changes at source.*

9.5 CHAPTER SUMMARY

We began our discussion of maintenance at the activity level and defined the terms used in maintenance. Planned work is more efficient than unplanned work and reactive work is less efficient as it is unplanned. Reactive or breakdown work is perfectly acceptable when the consequences of failure are small. In this case it is usually the lowest cost option. Therefore proactive work is not always strategically appropriate.

We examined the primary role or the raison d'être of maintenance. We developed a risk limitation model, using the escalation and damage limitation models discussed in Chapter 8. The viability and profitability of any organization, both in the short-term and in the long-term, are dependent on its ability to contain minor events and prevent them from escalating into major incidents. Far from being an interruption of production or an unavoidable cost, maintenance ensures that the revenue stream keeps flowing. It has therefore a very positive role, and is not merely an activity of fixing, finding, or anticipating the equipment failure.

We have discussed the continuous improvement cycle and its constituent maintenance phases. The objectives of planning are to achieve the required level of availability of the safety and production systems. Some breakdown or corrective work will result from the test routines and condition monitoring tasks. Breakdown work need not be entirely unplanned. We can prepare some generic plans to cover common breakdown types, with details being worked out when the breakdown occurs.

The objective of scheduling is to minimize production losses during the execution of the work. The first step is to prioritize the work, with safety at the top, followed by production, and then the rest. The scheduler arranges resources, tools, spare parts, logistics support, vendor assistance, permits, and other requirements for the safe and efficient execution of the work.

In the execution phase, safety is the first objective. Good planning and communication is essential for personnel as well as plant safety. Toolbox talks, the permit to work system, protective safety apparel, and proper planning are important steps in managing safety. Some jobs are inherently hazardous, requiring good planning and preparation to minimize the risks.

The second objective is to ensure that we do the work to acceptable quality standards. The quality of operations and maintenance affects the reliability of the equipment. The knowledge, skills, and motivation of the operations and maintenance crew will contribute to the quality of work.

The main drivers of maintenance costs are the operational reliability of the equipment and the productivity of the workforce. Quality is the result of the knowledge, skills, pride in work, and good team spirit of the workforce. It is the responsibility of management to create the right conditions to make these happen.

The third objective is to do the work in time. We examined the reasons for planned work not being done in time. As long as we do the work within an acceptable time span around the due date, we can consider the execution to be in compliance. When we complete the work outside this span, it is unacceptable. Compliance with the schedule is an important performance measure, and should be recorded.

Analysis is the last phase in the continuous improvement loop. We have to measure performance during each of the phases.

The fourth objective is to use resources efficiently. Good planning and scheduling are essential pre-requisites for obtaining high productivity levels. Delays caused in work are often attributable to poor scheduling.

We tend to overanalyze the execution phase so we need fewer, but more focused metrics. We should record safety and environmental incidents, as they are not acceptable and must be controlled. Quality is important and the trend in reliability is a good metric to use. For this purpose, we have to record the results of tests, the occurrence of spurious events, and equipment failure dates. It is also necessary to maintain compliance records. If we carry out the analysis work satisfactorily, it will result in improvements to the plan. If there are no improvements, either the plan was perfect or the analysis was inadequate. A qualitative measure of the analysis phase is therefore available.

We applied the risk-based approach to three situations commonly encountered in process plant maintenance.The examples included PRVs, duty/standby operation and end-to-end testing of control, loops.

References

1) Narayan, V. 1998. "The Raison d'être of Maintenance." Journal of Quality in Maintenance Engineering, vol. 4, no. 1. ISSN: 13552551. 38-50.
2) Leeson, N. 2000. Rogue Trader. Person ESL: ISBN: 058243050X
3) Deming, W. Edward. 2000. *Out of Crisis*. MIT Press. ISBN: 0262541157.

Appendix 9-1

A GENERALIZED VIEW OF MAINTENANCE

The approach we have developed so far could be considered for applications in other areas not related to equipment maintenence. Thus we can maintain law and order, the health of the population or the reputation of the business. There are risks to manage in each of these cases. We can use the event escalation model and the damage limitation model discussed in Chapter 8 in these situations as well.

In the case of law and order management, serious crimes such as murders and rapes are at the apex of the risk model. Similarly, the inability to manage unruly crowds can lead to situations that could result in fatalities. Soccer crowd violence has resulted in many incidents in the past few years, some of which resulted in multiple deaths. The law enforcement agencies have taken several steps to combat this problem. These include data banks with details of known trouble makers in several countries. The countries concerned share the data amongst themselves. Police from the guest nation travel to the host nation during major international matches to assist the latter in crowd management. They separate the fans from the guest nation physically from those of the host nation. They use closed circuit video cameras to monitor the behavior of the crowd. They control the sale of tickets carefully and route them through the soccer associations of the participating countries. They limit the sale of alcohol in the vicinity of the soccer stadium. For damage limitation, they use police on horseback, so that they can get in among the crowd and do so quickly. These are some of the *Plant, Procedure,* and *People* barriers that we use to prevent the escalation of minor disturbances into serious incidents.

Some of us will be familiar with the damage limitation procedures used when there is suspicion of food poisoning. If the authorities know the source, they quarantine the shop, factory, or warehouse. Countries can quarantine visitors traveling from a yellow fever area to prevent spread of this disease, if they suspect them to be carriers.

The case of the Barings Bank collapse[1] illustrates how financial risk management can fail, when the risk-control barriers are not in place.

The management of the reputation of a business has to deal with the qualitative aspects of risk. Here perception is reality, so concentrating on facts alone is not enough to convince people. People are more likely to believe firms who come clean when things go wrong than those who attempt to cover-up. Firms that have earned a reputation over the years of treating their people well will be better able to face difficult circumstances than others. Late in 1997, Levi Strauss (the maker of jeans) announced major staff cuts in the United States.[2] They had established a strong reputation for looking after their staff, and the public received the bad news very sympathetically. Even employees who lost their jobs had a good word for the company.

In 1982, seven people died after taking Tylenol, a pain-relieving drug made by Johnson & Johnson (J&J).[2] The authorities found evidence of tampering, and the drug was laced with cyanide. J&J immediately removed all 31 million bottles of Tylenol from stores and recalled purchases by customers. This cost them nearly a hundred million dollars, and was a vivid demonstration of its concern for the welfare of its customers. Tichy[2] notes that in 1975, J&J's chairman, James Burke wrote to all the senior managers, asking them to review their company Credo. Then he met hundreds of them in groups, and discussed their Credo. The opening line of their Credo emphasizes their responsibility to doctors, nurses, and patients. After much deliberation, the company as a whole decided to retain the Credo with only minor changes in wording. In the process, they had all committed themselves wholeheartedly to the Credo. James Burke credits these reviews to their success in handling the Tylenol crisis. He believes strongly that this large decentralized organization could not otherwise have managed the crisis. The public rewarded J&J by being loyal to it throughout the difficult times. Within three months, Tylenol had regained 95% of its market share.

In both cases, the public perception was that the firms had been honest and diligent. The policies they adopt in critical circumstances reflect their values.

There is an element of simplification in all these examples. Issues affecting people are always more complex than those affecting machinery. However, we can still apply the risk model.

Appendix References

1) Leeson, N. 2000. *Rogue Trader.* Pearson ESL. ISBN: 058243050X.
2) Tichy, N.M., and E.B. Cohen. 2000. *The Leadership Engine*. Diane Publishing Co. ISBN: 078816886X.

CHAPTER 10

Risk Reduction

In the preceding chapters, we examined hazards that we can expect during the lifetime of a process plant. The first step in managing them is to identify and, where possible, measure the risks. We can measure quantified risks using their component parameters, namely, the frequency and severity of the events. If the risk is qualitative, we note the factors affecting the perceptions. There is an element of simplification here, since quantitative risks can affect qualitative risks and vice-versa.

We have to bring these risks down to a level that society will be willing to tolerate, and at a cost that we can afford. Ideally, the best time to do this is while the plant is being designed. This does not always happen for reasons such as a lack of awareness, time, tools, resources, or skills. Often, the project team may get a performance bonus if they complete the project in time and within budget.The main risk they worry about can be that of the size of their bonus! Thus, their own agenda may conflict with that of life cycle risks facing thc plant.

In this chapter, we will discuss a selection of tools that are applicable in the design and operational phases of the plant. Of these, Reliability Centered Maintenance or RCM has a wide range of applicability in both the design and operational phases, so we will discuss it in some detail.

In Chapter 7, we explored the qualitative aspects of risk and why perceptions matter. If we are to influence the perceptions of the stake-holders in the business, whether they are employees, shareholders, union officials, pressure groups, or the public at large, we have to communicate our position effectively. If an individual or pressure group fights a large organization, the public sees them as David and Goliath respectively, and the organization faces a very difficult task.

10.1 FREQUENCY OR SEVERITY?

In managing quantified risks, we can attempt to reduce the frequency or the severity, or both. Risk in its quantitative sense was defined in Section 7.4 as,

Risk = Frequency x Severity, or
Risk = Probability x Consequence

In Figure 10.1 we can see a set of curves where the product of probability and consequence are constant. The risk may be in terms of loss of life or serious injury, financial loss, or damage to property or the environment. Let us say that we wish to move down from a high risk level such as the upper curve with a risk value of $20,000. One such point is where the probability is 0.5 and consequence is $40,000. Any point on the next curve has a risk level of $10,000, and so should be acceptable. We can lower the risks if we reduce one or both of the elements. It may be possible to lower the consequence while the probability remains the same. The vertical line represents this change. Alternatively, it may be possible to move horizontally, keeping the same consequence and reducing the probability. The figure illustrates these options, but there is no restriction to move parallel to the axes.

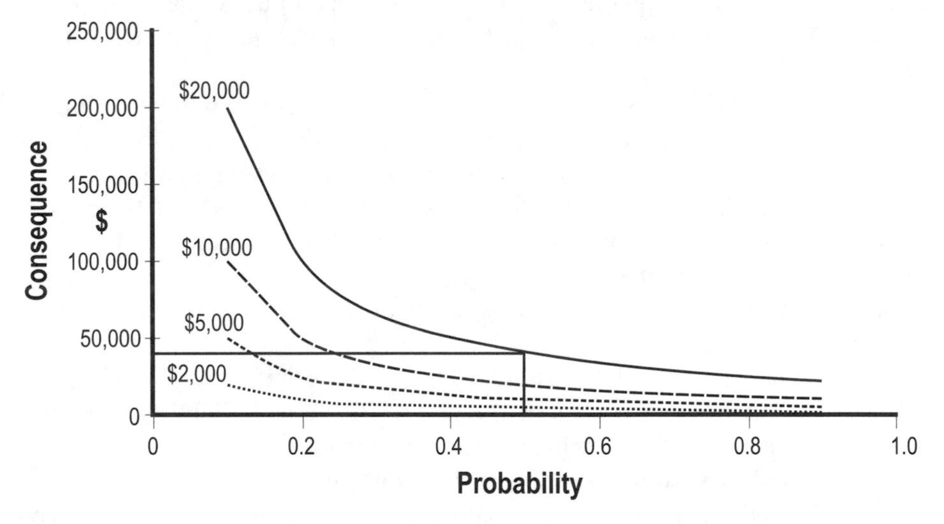

Figure 10.1 Constant risk curves.

In theory, both of these options are equally acceptable as they represent the same reduction in the risk value. However, people tend to accept high frequency events of low consequence, but they do not accept high consequence events of low frequency as easily. For example, a single road accident involving many vehicles and lives will catch media and public attention, but they are likely to account for a small proportion of the total road deaths on that day. Yet the remaining accidents are less likely to make it to the newspaper front page or TV headline news. In order to match our effort with people's perceptions of risk, it is preferable to look for the low consequence solutions. If the choice is between one or the other, we suggest risk reduction programs that mitigate the consequences in preference to those that attack the frequencies.

10.2 RELIABILITY BLOCK DIAGRAMS AND MATHEMATICAL MODELING

We introduced RBDs in Chapter 5, using series and parallel networks to represent simple systems. These represent the logic applicable to the physical configuration. In Figure 10.2, valves A and C isolate control valve B, and valve D bypasses it. The logical requirements for the flow to take place are that valves A, B, and C are all open or valve D is open. Thus, in the RBD, the blocks A, B, and C will be in series while the block D will be in parallel, as shown in fig 10.3.

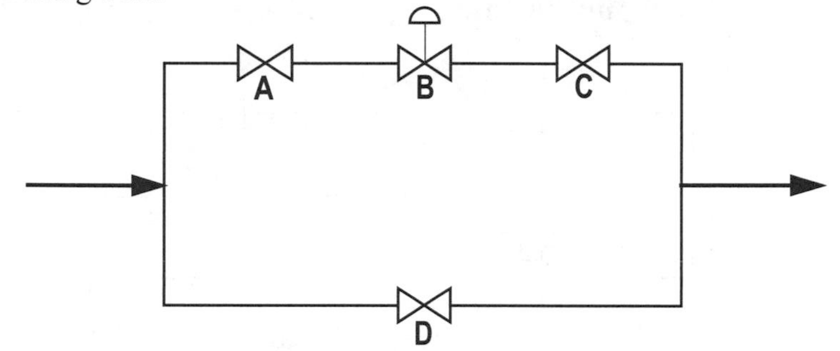

Figure 10.2 Physical layout of valves.

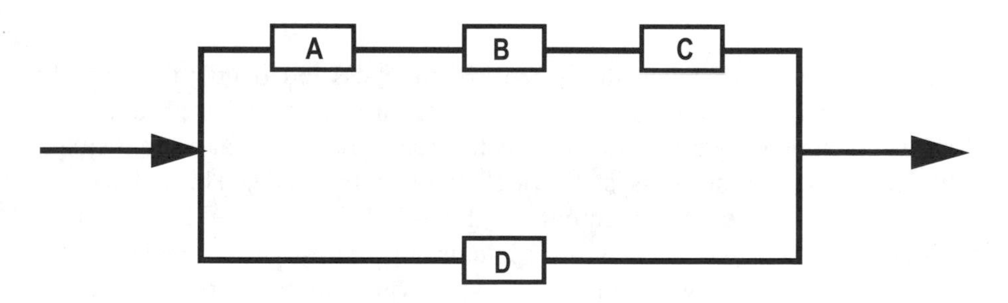

Figure 10.3 RBD of sub-system.

In Boolean algebra notation, we use AND gates to connect series blocks, and OR gates to connect parallel blocks. We ask the question whether both A and B have to operate in order to perform the function to decide how to represent them in the RBD. In the first case, the connection is with an AND gate and is a series link, while in the second there is an OR gate and a parallel link. The more complex arrangements include for example bridge structures or nested structures as shown in Figures 10.4 and 10.5 below.

The plant's overall system effectiveness is the ratio of the actual volumetric flow through the system to that possible when there are no constraints at the supply or delivery ends. It takes into account losses due to trips, planned and unplanned shutdowns, and slowdowns attributable to equipment failures. We factor in low demand or feedstock unavailability into the denominator. Thus

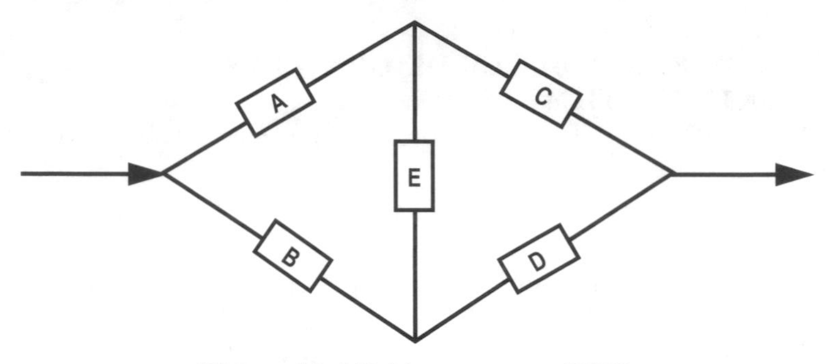

Figure 10.4 Bridge structure RBD.

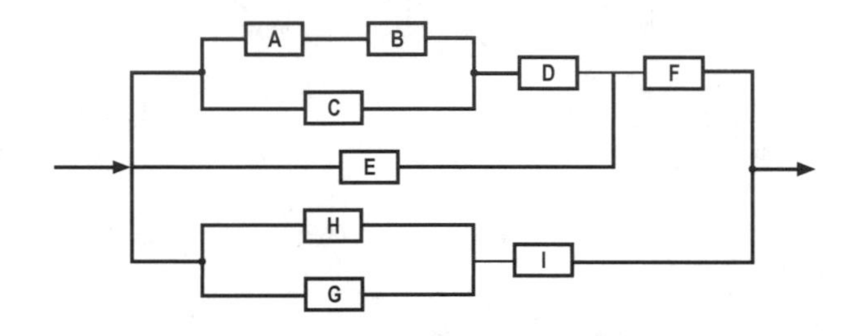

Figure 10.5 Series-parallel nested RBD.

we can use RBDs to evaluate the system effectiveness and identify the criticality of the individual blocks. Critical systems are those which produce the largest changes in overall system effectiveness when we make some small changes to each of the sub-systems or elements in the RBD. This helps us to carry out selective improvement of the reliability of the blocks concerned. The focus is on critical sub-systems only, whose operational reliability affects the overall system effectiveness. As a result, we can maximize the return on investment and net present value.

We can improve these sub-systems in a number of ways, for example,

- By changing the configuration;
- By providing storage for raw materials, and intermediate or finished products;
- By using more reliable components.

You will recall that availability is the ratio of the time an item can work to the time in service (refer to section 3.7). With a high equipment availability, the system effectiveness will also be higher. One way to increase availability is to improve the maintainability. We can do this, for example, by providing better access, built-in-testing, diagnostic aids, or better logistic support. When spare parts are available at short notice, this reduces downtime, which results in an improvement in both maintainability and availability. Last, as this is likely to be the most expensive option, we can consider installing standby equipment.

One can use analytical or simulation modeling techniques in the calculation of system effectiveness. Analytical solutions, using truth tables or algebraic equations, produce the same answer each time, and are deterministic. On the other hand, simulation methods, using numerical techniques, will produce different answers with each calculation. These differences are not errors, but represent the distribution or spread in the value of the outcome. Usually we require several simulation calculations, sometimes up to a thousand runs. We can produce a distribution curve similar to the pdf curve discussed in Chapter 3, by plotting all the results. Such results are probabilistic and give us the distribution, thereby offering a more realistic representation. Simulation methods give probable outcomes, while analytical methods produce possible outcomes.

Use of RBDs is particularly effective at the design stage, focusing attention on the critical parts of the system. The RBD is of use in achieving high system effectiveness and thus a low production risk. Different elements in the RBD will affect the overall system effectiveness differently. It is cost-efficient to improve those elements that produce the most overall improvement in system effectiveness. Once we establish the sensitivity of each path of the RBD, we check the sensitivity of the system as a whole to small changes in the parameters. In doing so, we start with the most sensitive equipment, then the next in order of sensitivity, and so on till we obtain the desired system effectiveness. These changes may be in the configuration, equipment reliability or capacity, storage at the supply, intermediate or delivery end, logistics, and installed spare capacity. We choose the combination of changes that produces the required improvement, that is, risk reduction, at the lowest cost. Similarly, we can question the need for some of the low sensitivity equipment, with potential savings in investment. This top-down approach focuses on items that will bring the greatest returns at the lowest cost, thereby making it a valuable decision tool.

Modeling is also useful during the operating phase. We can predict the effect of changes in e.g., spare parts unavailability, logistics support, trips or reliability at the equipment level on the system effectiveness and hence overall production capability. Again, we select the option that produces the greatest improvement in system effectiveness for a given cost.

A number of software applications are available to model systems, using analytical or simulation techniques. Simulation packages take longer to run, but require fewer assumptions. They can represent a wider range of real life constraints such as queuing for maintenance, varying demand, resources, and logistic support. They are useful for life-cycle cost evaluations, and may be applied from the conceptual stage of a project through the construction, commissioning, operating, and end-of-life phases of a plant. They are thus cradle-to-grave tools.

10.3 HAZARD AND OPERABILITY STUDIES

A Hazard and Operability study, or HAZOP, is a qualitative method of analyzing hazards and operability problems in new or change projects. It is a structured process to analyze the likelihood and consequences of initiating events. HAZOP uses a set of guide words to carry out the analysis. It is usually applied in turn to each element of the process. The team members allow their imagination to wander and try to think of as many ways in which they can expect operating problems or safety hazards, using the guide word as a directional prompt. For example, the guide word NONE will prompt the idea of no flow in a pipe line. In turn, this could be due to no feed-stock supply, failure of upstream pump, physical damage to the line, or some blockage. Other guide words such as MORE OF, LESS OF, PART OF, MORE THAN, and OTHER THAN will help generate ideas of different deviations that may cause a hazard or operability problem. They identify and record the consequence of each of these deviations. Corrective action required to overcome the problem, either by making the operator aware or by designing it out altogether. Such actions may involve additional hardware, changes in the operating procedures, materials of construction, physical layout, or alignment.

The HAZOP team should have a representative each from operations, process, and the mechanical and instrumentation engineering disciplines. A well-experienced and independent HAZOP team leader should facilitate the work of the team.

The technique helps identify environmental and safety hazards, as well as potential loss of production. It draws on the wealth of experience in the organization. By providing a structured approach, the team uses its energy efficiently. It is a pro-active tool suitable for use during the design phase of the project. Additional information on this technique is available in Kletz.[1]

10.4 FAULT TREE ANALYSIS (FTA)

A fault tree is a graphical representation of the relationship between the causes of failure and the system failure mode. Bell Telephone Laboratories Inc., introduced the technique in the early 1960s, and since then, it has grown in popularity. Designers use it to evaluate the risks in safety systems.

In the nomenclature of FTA, the TOP event is the system failure mode, while PRIME events are the causes. Table 10.1 describes a set of symbols used in constructing the FTA charts. We define the TOP event clearly by answering the questions what, when, and where. A TOP event, for example, is the loss of containment (what), during normal operation (when), from reactor R-301 (where). From this TOP event, we identify those causes that are necessary and clearly linked to it. Using the appropriate logic gate from Table 10.1, we can show the relationship between the TOP event and the immediate

(AND gate symbol)	AND Gate All inputs are required for output
(OR gate symbol)	OR Gate Any input will produce output
(rectangle symbol)	Fault event
(circle symbol)	Primary failure
(triangle symbol)	Transfer in/out
(diamond symbol)	Fault event not developed to cause

Table 10.1 Symbols used in fault tree analysis.

cause graphically. Next we identify the events that lead to these causes. This breakdown proceeds level by level, with all inputs to every gate being entered before proceeding further. We can stop the analysis at any level, depending on the degree of resolution required. We record the probability of occurrence of each of the causes, starting at the lowest level. Using the AND/OR logic information, we can calculate the probability of higher level events, ending with the TOP event. We can carry out what-if analysis, so that if the TOP event probability is unacceptable, the focus is on improvements to the critical branches. Figures 10.6 and 10.7 show a schematic drawing and an FTA chart respectively. For a more detailed explanation, readers may refer to Davidson[2] or Hoyland and Rausand.[3]

Software tools are available to construct FTAs and evaluate the probability of the TOP event. These lend themselves to sensitivity studies, and cost effective remedial measures. The use of FTA can reduce the probability of failure, and it is most appropriate at the design stage of a project. It is an analytical method and hence has the disadvantage of not being able to predict the spread or distribution of the results.

10.5 ROOT CAUSE ANALYSIS

We use this technique to improve reliability by identifying and eliminating the true reasons for a failure. The process is like peeling an onion, where the outer layers appear to be the cause but are effects of a deeper embedded reason. At

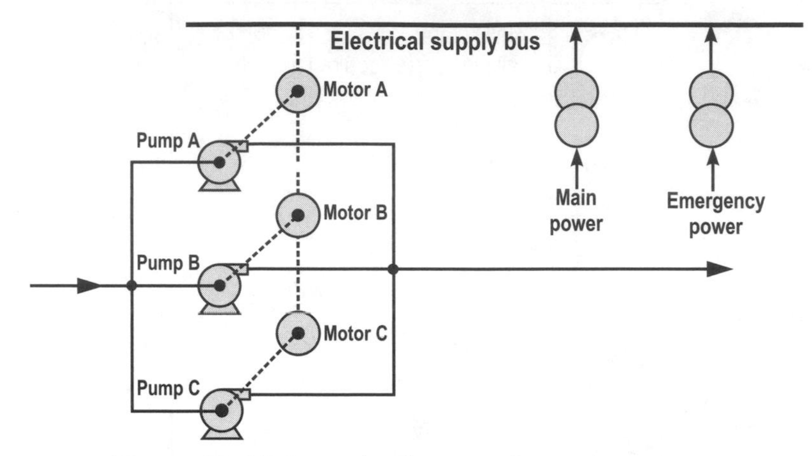

Figure 10.6 Schematic diagram of pumping system.

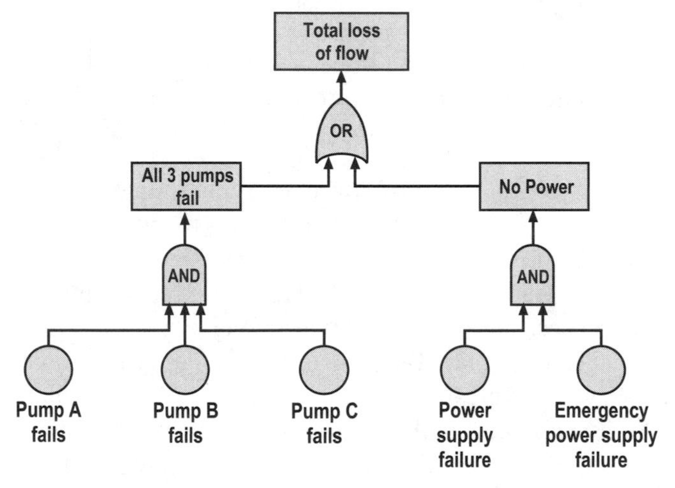

Figure 10.7 FTA of pumping system.

the plant level, for example, we may have visible problems such as environmental incidents, high costs, or low availability. On initial investigation, one may attribute it to poor logistics, high turnover of staff, absenteeism, or human error. Further investigation will reveal a variety of underlying layers of reasons, and one has to pursue it doggedly to arrive at the true cause.

We use a number of quality tools in carrying out root cause analysis (RCA). Many readers will be familiar with the Kepnor-Tregoe©[4] methodology. The change model and differentiation technique (Is, Is Not analysis), are powerful tools used in RCA (refer to Figure 10.8). The Fishbone (or Ishikawa) analysis technique, illustrated in Figure 10.9, helps identify proximate causes.

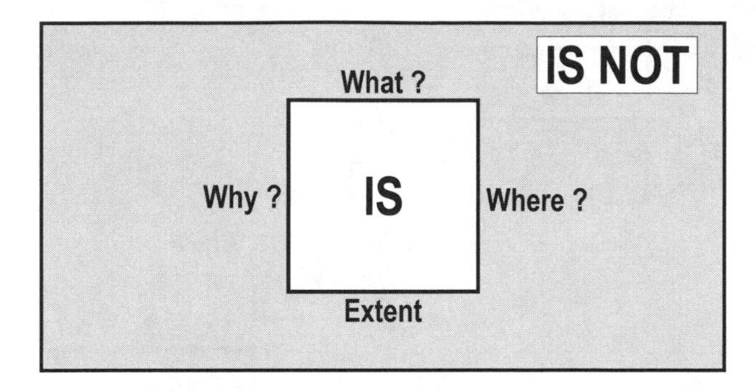

Figure 10.8 Problem solving models.

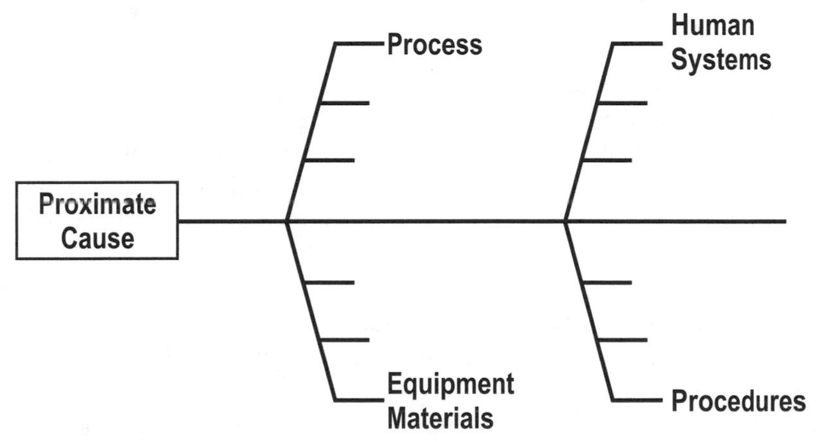

Figure 10.9 Fishbone (Ishikawa) analysis

Cause itself can be at many levels. If we are investigating a vehicle accident, we might consider several *possible* causes, for example, road condition, mechanical defects, or driver error. If the weather conditions were known to be bad, the poor road condition now becomes *plausible*. If we observe skid marks on the road, this evidence elevates it to a *proximate* cause. We examine which element of a proximate cause had the potential to do so, and look for supporting evidence. If such evidence is available, these potential causes become probable causes. We then test the most probable causes against the original effect or incident. If it can explain the full sequence of events, we call it the root cause. Figures 10.8, 10.9, and 10.10 illustrate some of the tools used in the analysis.

RCA is a structured process and the methodology is analogous to that used in solving a crime. We use problem statements (or inventory), classification, and differentiation techniques to identify the problem (crime). Thereafter, we describe the problem (evidence) using, for example, Pareto analysis, change model, timeline charts, and is-is not analysis. Next we identify the possible

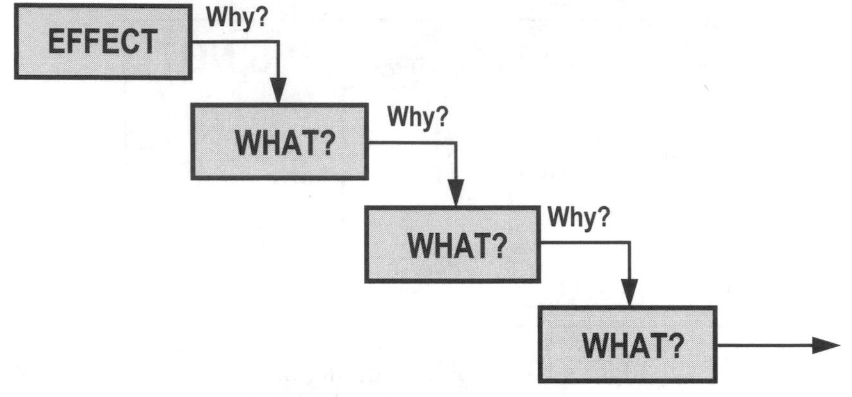

Figure 10.10 Stair-step analysis

causes (suspects) using the Fishbone and stair-step tools. Verification of plausible causes follows this step (confession, autopsy, trial). These steps result in identifying the most probable cause (judgment) that can explain the observed effects. Once we establish the root cause, the solutions are usually apparent. We can use tools such as brainstorming to find the best solutions (rehabilitation). An excellent example of the RCA technique is the report of the Court of Inquiry into the Piper Alpha disaster, conducted by Lord Cullen[5].

We reviewed some of the causes of human failures in Section 4.9. When an RCA indicates that human failure is a proximate cause, we have to peel more layers of the onion. It is necessary to find out if these are due to stress, lack of sleep, poor motivation, or other causes. Conflicting demands, such as pressures to keep the production going when safety is at stake, are not uncommon. We discussed the Piper Alpha disaster in some detail in Section 8.1.2. The rapid escalation of the fire was due to the large supply of hydrocarbons from Tartan and Claymore. Both continued to operate at full throughput when their Offshore Installation Managers (OIMs) knew that there was a serious incident at Piper Alpha[5]. They convinced themselves that Piper Alpha could isolate itself, when there was evidence that this was not happening. Swain and Guttmann[6] estimate that the probability of human error in well-trained people working under highly stressful conditions varies from 0.1 to 1.0. Even if we take the lower estimate, it is unacceptably large. Under these conditions, people tend to revert to their population stereotypes (doing what comes naturally). A strong safety culture in Tartan and Claymore may have persuaded their OIMs to shut down the pipelines connected to Piper Alpha. The evidence in Lord Cullen's report shows that this was not so, and the lack of the safety culture (population stereotype) was a contributor to the escalation of the disaster. The same scenario is evident in the Challenger shuttle case. The (production) pressure to launch was so high that both NASA and Morton Thiokol managers convinced themselves that the low ambient

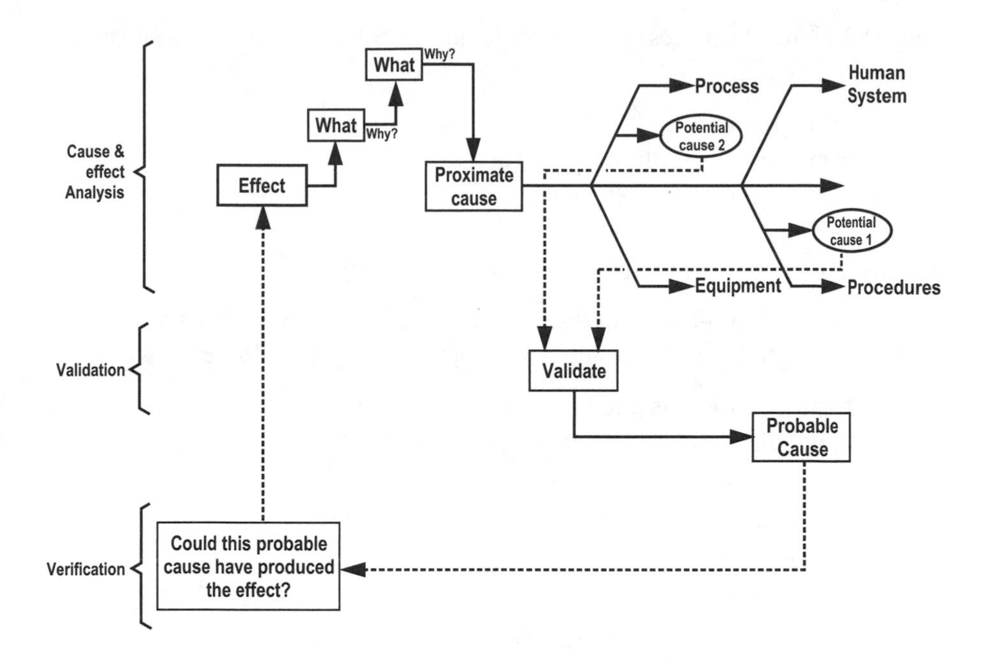

Figure 10.11 Root cause analysis steps

temperature did not matter. This was directly against the advice of their scientists. If there had been a strong safety culture in Tartan, Claymore, or NASA, the managers involved would in all probability have reached different conclusions. Successful RCA must be able to get to the underlying structural, emotional, and political pressures leading to human errors. We call these underlying causes *latent root causes*. These are sleeping tigers, waiting to strike. An RCA is complete only when the physical, human, and latent root causes are identified clearly.

RCA works by eliminating the source of problems, and thus improves the operational reliability of the plant, system, or equipment. As a result, we can expect a lower frequency of failures.

10.6 TOTAL PRODUCTIVE MAINTENANCE

In the early 1970s, Seiichi Nakajima pioneered the concept of total productive maintenance or TPM in Japan. The operator and maintainer form a team to maximize the effectiveness of the assets that they own. TPM embodies continuous improvement and care of assets to ensure that their operation at optimum efficiency becomes an organizational value driver. The operator applies five principles:

- *seiri* or being organized ;
- *seiton* or being disciplined and orderly;
- *seiso* or keeping the asset clean;
- *seiketsu* or remaining clean;
- *shitsuke* or practicing discipline.

TPM, which originated in the manufacturing industry, follows a five-step improvement path. The first step is to recognize six types of losses, as follows:

- unplanned shutdowns and breakdowns;
- additional changeovers or setups;
- trips;
- slowdowns;
- start-up losses;
- re-work and poor quality.

We can analyze these, for example, by using FTA and eliminating or minimizing the causes to the extent possible.

The second step involves the routine upkeep of the asset, by cleaning, condition monitoring, servicing, and preventive maintenance. The third step requires the operator to understand the importance of the machine quality in delivering the product quality. In TPM terms, we call this autonomous maintenance. The enhancement of the skills of the operator by training, both off and on the job, is the fourth step. The last step relates to designing out maintenance of the machine to the extent possible. Details on the methodology and application are available in Willmott[7].

10.7 RELIABILITY CENTERED MAINTENANCE

Reliability Centered Maintenance or RCM is a structured methodology to determine the appropriate maintenance work to carry out on an asset in its current operating context so that it performs its function satisfactorily. RCM identifies the timing and content of the maintenance tasks that will prevent or mitigate the consequence of failures.

We started our discussion on the RCM process in chapter 2, with an explanation of FBD and FMEA. In chapter 3, we explained the concepts of probability density functions, hazard rates and mean availability. Later, in chapter 4, we discussed the operating context, capability and expectation and incipiency. In this section, we will go through the whole RCM process. This is, in essence, a set of sequential tasks to identify the correct maintenance required to mitigate against the consequences of all credible failure modes. Further, using the knowledge acquired in chapter 4, we can determine the timing of these tasks.

We had set out to answer **what** maintenance to do and **when** to do it; these answers should be available by the end of this section. Following this brief discussion, readers are encouraged to refer to other texts on RCM (see bibliography) for a more detailed explanation.

10.7.1 Functional block diagrams

We discussed the functional approach in Chapter 2 and used FBDs to define the functions of the system and sub-systems, showing the inter-links that exist between them. The functional approach works in a top-down manner and identifies what each system or sub-system must achieve. From this, we define failure as inability to perform the function. It is a black box approach where raw materials or other inputs enter one side of the box, and intermediate or final products exit from the opposite side. The first two steps in an RCM study identify the functions and functional failures.

10.7.2 Failure modes and effects analysis

There can be a number of reasons that cause a functional failure, so the next step is to identify these. For example, if the discharge flow or pressure of an operating pump drops to an unacceptable level, there may be one or more causes. One is the blockage of the suction strainer, another an increase in the internal clearances due to wear. We call these causes failure modes and identify them by a local effect, such as an alarm light coming on or by a fall in the pressure or flow reading. In our example, the local effect is the drop in discharge pressure. This is how we know that something unwanted has happened. We use the human senses or process instruments in the control room or at site to identify the failures.

The failure may affect the system as a whole resulting in, for example, a safety or environmental incident, or production loss. In the above example, if there is an installed spare pump that cuts in, there is no loss of system performance. On the other hand, if this is the only pump available, the system will not function, causing a loss of production, or impairing plant safety. This is the system effect.

We then examine the category or type of consequence. It can be a hidden failure, as in the case of a standby pump failing to start or a pressure relief valve failing to lift. You will recall from the discussion in Chapter 4 that a distinct feature of a hidden failure is that unless there is a second failure, there is no consequence. This second or other event may be a sudden increase in pressure, or the failure of another equipment. Thus the operator cannot know that the standby pump will not start or that the relief valve will not lift unless there is a demand on the item. This happens if the duty pump stops or the pressure has risen above the relief valve set-pressure. By then it is too late, as we want the equipment to work when required. The effect could be impairment of safety, potential damage to the environment, or loss of production.

The failure may be evident to the operator in the normal course of duty. For example, when the running pump stops, the operator will know this by observing the local or panel instruments. If it is an important function, there will be an alarm to draw the attention of the operator.

The consequences of failures depend on the service, the configuration, and the external environment, and whether they are evident or hidden. We will illustrate this by examining a number of scenarios. First, they can have safety or environmental consequences, for example, when a pump or compressor seal leaks and releases flammable or toxic fluids. Even benign fluids may form pools on the floor, causing slipping hazards, and resulting in a safety consequence. Second, they may result in a loss of production. If a pump bearing seizes, the pump will stop and there will be no flow. If it does not have an installed stand-by unit, there will be no flow in the system, impairing safety or production. If there is a stand-by unit and it cuts in, the system continues to function. However, in this case the seizure can result in the shaft being welded to the bearing, so we may be in for a costly and time-consuming repair effort. Alternately, the seizure may result in internal parts rubbing, thereby causing extensive damage. Thus, the third consequence is an increase in maintenance cost. In this case, even though there is no impact on safety, environment, or production, there may still be a high cost penalty. Finally, there may be no effect at all on the system, in which case the failure does not matter. Taking the example of the bearing seizure, the result can be just damage to the bearing itself and nothing else. In this case, assuming that the cost of replacing it is small, we would classify the failure as one that does not matter. Categorizing system effects assists us in determining the effort we are willing to spend and, hence, the appropriate maintenance task.

These steps complete the Failure Modes and Effects Analysis or FMEA.

10.7.3 Failure characteristic analysis (FCA)

We discussed failure distributions, hazard rates, and failure patterns in Chapter 3. We also examined the special case of constant failure rates. When dealing with hidden failures, testing is an applicable and effective maintenance task. Often it is the most cost effective task. Under the conditions discussed in section 3.7, we can use expression 3.13 to determine the test intervals that will ensure the required level of availability for a given failure rate.

In the case of evident failures that exhibit incipiency, the time interval from incipiency to functional failure is of interest. Figure 4.10 in Chapter 4 shows a typical incipiency curve. The curves (refer to Figure 10.12) start randomly along the time axis, so the operator will not know the starting point of the incipiency. In order to measure a deterioration in performance, we need at least two points on the curve, so that we can recognize that performance deterioration has commenced. At what frequency should we test the item in order to ensure that we get at least two points on the curve? Let us choose a test interval equal to the incipiency period. As can be seen from Figure 10.13, it will be impossible to find two points on the curve within the incipiency period. If we choose a test interval of, say, two-thirds of the incipiency, we notice from Figure 10.14 that we will miss some functional failures. However, if we choose a test interval of half the incipiency, we will always get two points on the curve within the incipiency period, as illustrated in Figure 10.15. Thus, for evident incipient failures, the test interval cannot exceed half the incipiency period. In the case of safety or environmental consequences, we can select a smaller test interval, say one third the incipiency period.

We have to provide for the variability in the droop of the incipiency curve, as illustrated in Figure 4.14 in Chapter 4.

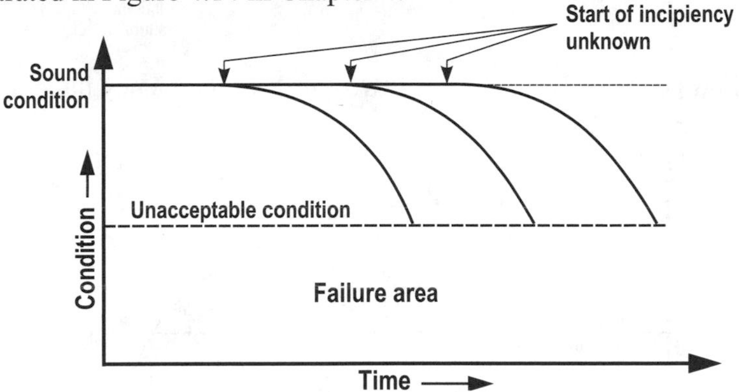

Figure 10.12 Incipiency curves with random starting points of deterioration.

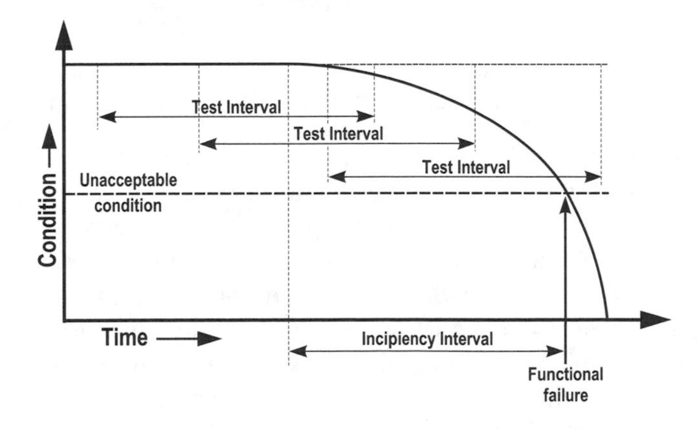

Figure 10.13 Showing why the test interval cannot be equal to the incipiency interval.

The analysis of failures is not as complicated as it may seem. Often, a discussion with the operators and maintainers will yield good reliability information. However, the analyst must ask the right questions in an unbiased way. If we have equipment run-times and special graphpaper, it is not particu-

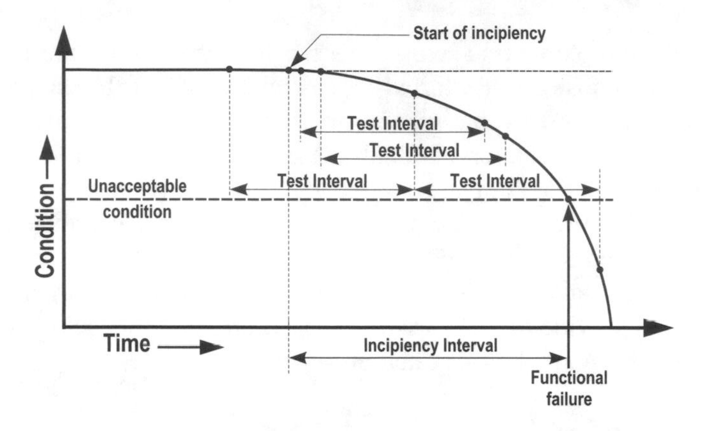

Figure 10.14 Test interval at two-thirds of incipiency interval.

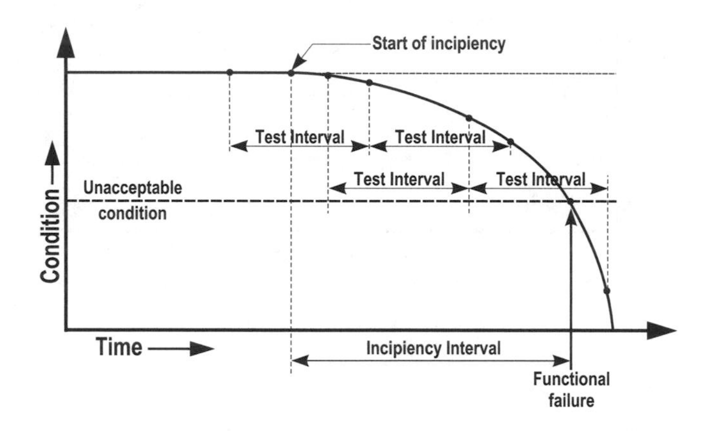

Figure 10.15 Test interval selected at half of incipiency interval.

larly difficult to carry out reliability analysis. We need access to the operating history, specifically the start-stop dates and the cause of stoppage on each occasion. Similarly, calculation of average failure rates requires the time in operation and the number of failures. Operators can usually identify incipiency intervals from their knowledge of the equipment performance.

For various reasons, some RCM practitioners shy away from reliability analysis. As a result, this is often the weakest area in the RCM process. A legitimate reason is the total lack of operating history, as in the case of a new plant. Another reason is that good quality data is often missing, as discussed in Section 3.9—the Resnikoff conundrum. It is true that data about really important failures is hard to find because of the great deal of effort spent in preventing such failures. In these cases, it is perfectly acceptable to use vendor recommendations or one's own experience as a starting point. Thereafter, we use age-exploration to refine the intervals based on the condition of the equipment when inspected, after operating it for some time. Each new inspection record adds to this knowledge, and using these we make further adjustments to the maintenance intervals. We make these changes in small steps, and check how they affect the equipment's performance. Smith[8] explains this method in greater detail.

However, some practitioners take the view that it is too difficult or costly to collect reliability data. They stop on completing the FMEA, even when data is available to calculate the reliability parameters. Instead, they use guess-work (euphemistically called engineering judgment), to determine the task frequency. We do not support this approach, since it devalues the R in RCM, and produces results of indifferent quality.

A large volume of maintenance work relates to minor failures on which data is available. Serious failures result from the escalation of minor failures, as discussed in Chapter 9. If we resolve minor failures in time, there is a good chance of avoiding serious ones. Continuous improvement is only possible when we collect and analyze performance data, so it follows that data collection is an integral part of good practice

10.7.4 Applicable and effective maintenance tasks

We discussed applicability of tasks based on the shape of the pdf curve. To illustrate this, we used the Weibull shape and scale parameters to help identify applicable maintenance tasks (refer to secton 9.1.2). There are other criteria to consider as well, and we will examine these now.

Effective tasks relate to the type of failure and those that will solve the underlying problem. We ask the following questions about the failure:

- Is it evident or hidden?
- Is there an incipiency and can we measure it?
- What is the hazard pattern (for example, infant mortality, constant hazard rate, or wear-out)?

The tasks must relate to the answers we get to these questions. For example, if there is a hidden failure, a failure-finding task is applicable. With a wear-out pattern, age-related maintenance (scheduled overhaul or replacement) is applicable. There can be one or more applicable tasks that can address a given hazard pattern and failure type (that is, hidden or evident).

Effective tasks are those that are technically capable of solving the problem. A person suffering from a headache can take a pain-killer tablet to obtain relief. Often applying balm or massaging the head will do just as well, so these two methods can also be effective. However, applying balm to the feet will not be useful, so this not an effective solution. What is important is that the task must address the cause of the problem.

An automobile tire wears out in the course of time. Failure of the tire has the potential for serious consequences, including loss of life. With a clear wear out failure pattern, a time-based replacement strategy can be effective. Alternatively, we can measure the tread depth from time to time, and plot an incipiency curve. Hence, condition monitoring is also effective. Both tasks are physically possible, so the strategies are equally applicable. What is the best option?

In the case of time-based tasks, such as failure-finding, scheduled overhaul (or replacement) work, we have to select its timing so that the residual life or survival probability at that point is reasonably high. The task will only be effective if executed in time. In other words, we cannot delay the task to the point where there is a good chance that the item has already failed—the patient must be alive to administer the treatment! Referring to Figure 3.2, the R(t) value is what we have to ensure is acceptable. In the case of safety and environmental consequences, we expect to see a high survival probability, say 97.5% or higher. The actual value will depend on the severity of the consequences, and may be as high as 99.9%. If we had selected a desired survival probability or R(t) of 97.5%, we expect that at this time, 2.5% of the items would have failed. For a single item, there is a 97.5% chance that the item would still be working when we schedule the maintenance task. We call this the **safe life**. Normally, we can accept a lower survival probability when dealing with operational consequences. The exact value will depend on the potential loss and could be as low as 85%. This means that we can delay the maintenance work to a point much later than the safe life, called the **useful life**. The following examples illustrate the concept of safe and useful lives.

Gas turbines provide motive power for modern aircraft. The heat produced in the burners expands the combustion products and imparts this energy to the rotor blades. The combustion takes place in several burners mounted uniformly around the inlet face. In theory, the hot gases must be at the same temperature at any point around the circumference, along a given axial plane. In practice there are small variations, as a result of which the blades see varying temperatures as they rotate. This causes thermal cycling and fatigue. The blades experience mechanical stresses due to dynamic centrifugal forces as well as due to the differential pressure between inlet and outlet. They fail due to these mechanical and thermal stresses, and can break off and cause extensive damage inside the casing. They can also burst out through the casing, causing injury to people or damage to property. Catastrophic blade failures are unacceptable, and we must do all we can to prevent them. The manufacturers test samples of these blades to destruction under controlled conditions, and assess their failure distribution. With this information, the manufacturers can predict the survival probability of the blades at a given age. The gas turbine manufacturers will recommend that the user re-blades the rotor at a very conservative age, when the blade survival probability is high. We call this the safe life of the blade.

Similarly, ball bearing manufacturers test samples of their products to destruction and plot their failure distributions. In the majority of applications, failure of ball bearings will not cause catastrophic safety or environmental incidents. In such cases, we can tolerate a lower survival probability at the time we decide to carry out maintenance action. Usually, we design ball bearing applications on the basis of its so-called L(10) life, which is the age at which 10% of the bearings in the sample tested by the manufacturer have failed. At this point in time, we expect to see 90% of the sample to have survived, a level far less stringent than in the case of the gas turbine blades. The bearing L(10) life is its useful life.

In Chapter 3, we discussed how the Weibull shape parameter describes the peakiness of the ***pdf*** curve. In Figure 3.16, we can see the shape of the ***pdf*** curves for Weibull β values from 0.5 to 10. The higher this number, the more certain we can be that the failure will take place close to the characteristic life. When the value of the shape parameter is low, for example 1.1 or less, the spread is very large.

In the case of evident failures, when we can be reasonably sure of the failure interval, time-based maintenance is applicable. With high Weibull β values, typically over 4, time-based maintenance can be quite effective. We can expect high Weibull β values with items subjected to fouling, for example, furnaces or heat exchangers. We can also expect high Weibull β values in items subjected to wear, such as brake pads or tires. Some people call time-based tasks hard-time tasks, scheduled overhauls, or scheduled replacements. Note that the word time encompasses any parameter that is appropriate in measuring age. We can replace it by start cycles or the number of operations if these are more appropriate in a given situation.

When the shape parameter is around 1, (say a range of 0.9 to 1.1), the time of failure is difficult to predict due to the wide spread of the probability density curve. If there is an incipient condition that we can monitor, then condition-based maintenance is a good option.

When the Weibull β value is lower than 1, say 0.9 or less, the item is subject to early failures or infant mortality. Here, the probability of failure decreases with age, so if the item has not yet failed, keeping it running is the best option. If we stop the item to do preventive maintenance, we are likely to worsen the situation. The low Weibull β value normally indicates a situation where the stress reduces with age, as in the case of internal parts that adjust and align themselves during operation. We use the term *bedding-in* to describe this learning process. Items such as crankshafts that have sleeve bearings, pistons, and gear trains, which align themselves after running-in for a few hours, illustrate this process. A low Weibull β value can also indicate quality problems, either in terms of materials of construction, maintenance workmanship, or poor operational procedures. In this case, a root cause analysis will help identify the underlying problems.

In the case of hidden failures, by definition we do not know the exact time of failure. If the item is mass produced, we can test a representative sample to destruction. Such a test, typically carried out on items such as switches or relays, helps establish the failure rate of the items. Often such tests are not practical, if the unit costs are high and testing to destruction is not viable. Similarly, when factory test conditions cannot match the operating scenario, they lose credibility. An alternate method is to test the item periodically in service, without adjusting, cleaning or modifying it in any way. If there are a number of similar items in service, we can calculate the average failure rate by dividing the number of failures by the cumulative operating service life in the test interval.

The failures we are talking about in this context are those recorded in the FMEA. For example, a pressure relief valve (PRV) may fail to lift at the required pressure, or may fail by leaking even when the system pressure is lower than the set pressure of the PRV. If overpressure protection is under consideration, we must base the failure rate on test results for these events alone. When we hear of a PRV failing in service, often it means that it is leaking. If so, this data is not relevant in calculating the failure rate relating to the fail-to-lift scenario. What data do we need for this calculation? The results of pre-overhaul bench tests will identify the number of fail-to-safe (leakage), and fail-to-danger (not lifting) events. The latter are relevant in this case and should be used to calculate the failure rate. We can calculate the frequency of future tests using expression 3.13. We call such tests failure-finding tasks as they identify the ability or inability of the item to perform when required. Expression 3.13 is applicable when we can fulfill the conditions mentioned in Section 3.8. If not, we can use a numerical method called the Maximum Likelihood Estimator or MLE. Edwards[9] describes the method in detail. When the failure event has no consequence or if it is very small, we can allow the equipment to run to failure. The item must fail before we do any maintenance work. A surprisingly large number of failure modes can fall in this category. With this knowledge, we can reduce the preventive maintenance work load significantly. Often such unnecessary maintenance results in additional failures caused by poor workmanship or materials. Eliminating the unnecessary maintenance will help reduce early failures, thus eliminating some breakdown work as well. The uptime or availability of the equipment also rises correspondingly.

Finally we have the situation where the failure matters, but we cannot find a suitable maintenance task that will mitigate the consequence. If the failure has a safety or environmental consequence, we have no choice but to redesign the system. In this case, we improve the intrinsic reliability of the system, so that the failure rate drops to a tolerably low level. We do not need to restrict such redesign to that of equipment. We have discussed the importance of people, procedures, and plant in Chapter 9. Training to raise the competence of people is a form of redesign. Similarly, revising the operating and maintenance procedures to reduce the failures is also a form of redesign.

When we carry out RCM studies, in about 5% or so of the failure modes we are usually unable to find an applicable and effective strategy. In such cases, if the failure affects safety or the environment, then redesign is the only available option. Applying RCM in new or change projects helps identify these failure modes while the design is still on the screen or drawing board. We can do such redesign work at relatively low cost and with minimum impact on the project schedule.

10.7.5 Cost-effective maintenance tasks

We noted earlier that there may be several applicable and effective tasks available to tackle a given failure mode. For example, one may test a smoke detector or simply replace it with a pre-tested unit. We can test items removed from the plant later in a workshop. In some cases, this procedure can be cheaper than testing at site, especially if downtime is expensive or otherwise unacceptable. In this case we replace failure-finding activity with a scheduled replacement task. In the case of oil or fuel filters, we need to clean or replace the choked elements. We can measure the onset and incipiency of failure by measuring the differential pressure across the filters. Hence an on-condition maintenance task is applicable and effective. If the rate of fouling is very predictable, a scheduled replacement task is also applicable and effective. In this case, the latter strategy can be cheaper as the timing will be the same as with condition monitoring, without incurring the cost of the latter. Sometimes there are convenient windows of opportunity to carry out maintenance tasks. For example, a gas turbine may be down for a scheduled water-wash. There may be a number of maintcnancc tasks on thc unit and its ancillaries for which failure-finding, condition-based, or on-failure strategies are applicable. However, we can reduce equipment downtime if we arrange to do these tasks during the water-wash outage.

The RCM logic requires us to find the most applicable and effective task in all cases. This is especially important in the case of failure modes that have a safety or environmental consequence. There may be more than one applicable task from which we select the best one. In the case of safety or environmental consequences, we select the most applicable and effective task as cost is of secondary importance. In the case of the tire wear out situation, a time-based or mileage-based replacement can result in some tires being replaced too early, as they are not fully worn out. Similarly, some other tires may have exceeded the tolerable wear out limits and may pose a safety hazard. Hence this strategy is not optimal. If we replaced tires based on the tread depth, we can be sure that it is replaced at the right time. It is the most cost-effective task, as the monitoring is inexpensive and we do not replace the tires prematurely. So in this case the condition-based strategy is the most cost-effective one to use.

10.7.6 Task selection

In RCM terms, we apply strategy at the failure mode level. We have discussed two essential criteria in selecting strategy, namely, applicability and effectiveness. Table 10.2 shows these criteria and how they influence the strategy selection. The actual task selected will depend on the operational context. In the case of safety and environmental consequences, we select the task that will reduce the risk to a tolerable level. With operational and non-operational consequences, we select the most cost-effective solution. Table 10.2 shows a list of applicable and effective strategies for the different scenarios. We have to judge their cost-effectiveness on a case-by-case basis.

10.7.7 Preventive maintenance routines

Once we find suitable tasks for all the failure modes, we can start writing the preventive maintenance routines. In order to minimize equipment downtime, and optimize the utilization of resources, we propose the following steps:

- Sort the tasks by frequency, for example, as weekly, monthly, or annual;
- Sort the tasks by main equipment tag, for example, pump P 4010A;
- Sort the tasks by main resource, for example, mechanical technician;
- Examine the sorted lists and rationalize;
- Examine whether a higher level task can replace a number of tasks addressing individual failure modes. For example, fire regulations often require us to start up emergency diesel-engine driven fire-pumps on a weekly or fortnightly basis. If the test is unsuccessful, we can then look for weak batteries, damaged starting motor or clutch, fouled air or fuel filters, blocked fuel injectors, or a worn fuel pump. There is no need to carry out the latter tasks on a preventative basis. The maintenance routines will then be a weekly start of the equipment;
- Create the maintenance routines using judgment, local plant knowledge, and experience.

Software packages are available to help carry out RCM analysis. These can vary from the highly configurable package, which has very little RCM logic embedded in it, to the highly structured one that require mountains of data entry for each failure mode. It is better to use RCM packages that have the logic built-in and do not require large volumes of input data. User friendliness is important; we want RCM specialists with some computer literacy to do the job, and not require computer specialists. Software packages have many advantages, including speed of execution, audit trails, and quick search and retrieval facilities. We can create child studies from an original parent-study, noting relevant changes in operating context and modifying the FMEA, FCA, and strategies suitably.

Paper-based systems have some disadvantages. It is difficult to keep them up to date. Users do not have built-in check lists and cannot trap errors easily. It is also difficult to search for data, and using them is labor-intensive. Software based systems overcome these problems; they facilitate easy data exchange, and can be customized.

Consequence	Evident or Hidden	Failure Timing Predictability			Condition Monitoring or Testing Possible?	Applicable Strategies	Selection Criteria
		Well defined	Constant hazard	Wear in			
Safety or environmental	Evident		Yes		Yes	On-Condition	Tolerable risk
	Evident	Yes				Scheduled (safe life)	Tolerable risk
	Evident			Yes		Re-design	Other strategies ineffective
Safety or environmental	Hidden	Yes				Scheduled (safe life)	Tolerable risk
	Hidden		Yes			Failure finding	Tolerable risk
	Hidden			Yes		Re-design	Other strategies ineffective
Operational	Evident		Yes		Yes	On condition	Cost-effective
	Evident	Yes				Scheduled (useful life)	Cost-effective
	Evident		Yes			On failure	Cost-effective
	Evident			Yes*		Re-design	Other strategies ineffective
Operational	Hidden		Yes		Yes	Failure Finding	Cost-effective
	Hidden	Yes				Scheduled (useful life)	Cost-effective
	Hidden		Yes			Failure finding	Cost-effective
	Hidden		Yes			On failure	Cost-effective
	Hidden			Yes*		Re-design	Other strategies ineffective
Non-operational	Evident				Yes	On condition	Cost-effective
	Evident	Yes				Scheduled (useful life)	Cost-effective
	Evident		Yes			On failure	Cost-effective
	Evident			Yes*		Re-design	Other strategies ineffective

Table 10.2 Applicable Maintenance Strategies

*If the wear-in period is clearly related to bedding-in, take no action.

RCM is an excellent tool that successfully reduces risk, but it is an expensive and time-consuming process. It is equally effective in safety, environmental, production, or cost critical areas. Smith[8] quotes a 1992 report by the (U.S.) Electrical Power Research Institute study, in which the average pay-back period for the early RCM work in the utilities is 6.6 years. He expects this to reduce to two years for more mature work. The real benefit comes from the increase in plant availability and, hence, improved integrity and profitability.

We carry out RCM analysis on systems that pose the greatest risk—to safety, the environment, and production capability. The selected systems will normally account for about 20-30% of the maintenance workload in a plant. Figure 10.16 shows a selection chart to help identify critical systems. Once we select the minimum number of systems to work on, it is advisable not to take short-cuts in the procedure itself. Templating or copying projects without adequate consideration to the operating context and physical similarity of equipment may appear to save money. Actually, it wastes time and money because it will not produce technically acceptable results.

10.7.8 Structural and Zonal RCM Analysis

So far we have looked at RCM Systems Analysis. Two other RCM methods which do not use the FMEA approach are Structural and Zonal Analysis. They are risk based and cost effective. A detailed discussion on these techniques is outside the scope of this book, but readers may wish to refer to Nowlan and Heap[14] for additional information.

10.8 COMPLIANCE AND RISK

We had defined planning as the process of thinking through the steps involved in executing work. This process helps identify the risks. With this information, we can find ways to reduce these risks to a tolerable level. In this chapter, we have looked at a number of tools that can help us to reduce risk effectively. Once we identify and schedule the right work, we have to follow through and execute it in time to the right quality standards.

We discussed the use of compliance bands in Section 9.3.3. The manager can alter the width of these bands to suit the circumstances in a given plant. If we complete all the jobs on the scheduled date or within the agreed band, the compliance is 100%. In practice, it is likely to be lower than 100 % due to equipment or resource unavailability, or due to market constraints. Such non-compliance increases the risk of safety and environmental incidents as well as potential loss of production. Referring to Figure 10.1 on risk contours, we are in effect moving from a lower-risk curve to one that is higher.

10.9 REDUCING PERCEIVED RISKS

Perceptions are not easy to handle because we do not always know the underlying reasons, and they do not follow any simple structure or logic. Often, people do not express their feelings and emotions so we may not even be aware of their existence. Nevertheless, we can and should reduce these risks to the extent possible. Good communication with the stake-holders is important, something that is easier said than done.

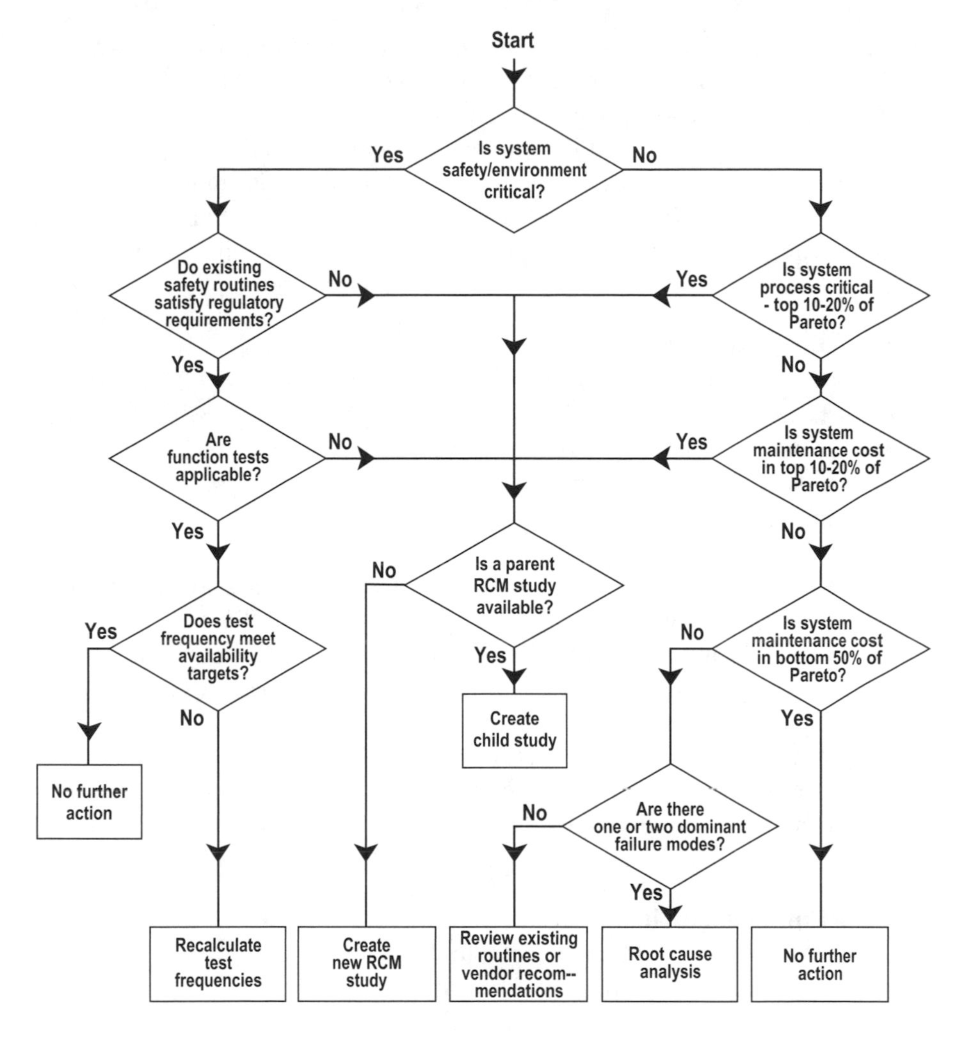

Figure 10.16 RCM selection criteria.

10.9.1 David and Goliath scenarios

In the mid-1980s, two unemployed environmental activists attacked McDonald's, the world's biggest restaurant chain. The activists were associated with London Greenpeace (not connected with Greenpeace International). Their leaflets criticized McDonald's record on health, the environment, animal rights, and labor relations[10,11,12,13]. McDonald's sued, and the case lasted 213 days over a three year period. It was the longest libel action in England. The judge ruled that the defendants' statements injured the plaintiff's reputation. However, he agreed with the defendants that the company's advertisements were exploitative of children, that it was responsible for cruelty to some animals and that it paid low wages[10].

The case drew a great deal of media and public attention. There was a support web site and a so-called McLibel Support Campaign. At the 1995 shareholder's meeting in Chicago on May 26, 1995, there were repeated questions to Michael Quinlan, the Chief Executive whether it was in the company's best interests to continue the suit[13]. He replied that the case "is coming to a wrap soon." In the event the case carried on till June 26, 1997. Whether they are right or wrong, as far as the public is concerned, Goliath is guilty. A court victory need not result in changing the hearts and minds of the public.

10.9.2 Influence of perceptions

We discussed the factors that affect perceptions of risk in Section 7.2. These perceptions have an influence on the way we make decisions. *Two of these factors—dread and fear of the unknown—have a particularly strong influence.* Poor communication contributes to both these factors, so we have to address this with some urgency. If the plant or facility is close to a populated area, it is important to carry on a dialogue with those who live there. One must use care and tact so that one does not raise unnecessary fears. The intention always is to reduce the fear of the unknown, while not creating a sense of dread.

We must communicate emergency response plans to the people in the surrounding areas and coordinate these with those of other facilities in the vicinity. We must work out these plans in consultation with the community. If there has been a near-miss that has the potential to harm the neighborhood, prompt disclosure will help improve credibility. If members of the community visit the facility periodically, they can see for themselves how the plant manages environmental and safety issues.

10.9.3 Public goodwill

In the case of Johnson & Johnson (refer to Appendix 9-1), the Tylenol disaster was so well managed that they got the public on their side. After the event, sales rebounded and J&J continued to prosper. The company had to earn the goodwill, and this does not happen overnight.

10.10 CHAPTER SUMMARY

Managing risks requires that we understand and find effective means to reduce them to a tolerable level at an affordable cost. The best time to do this is while the plant is being designed. In this chapter, we have looked at some of the issues that are relevant.

The qualitative aspects of risk are important and perceptions matter. In addressing risks, we have to take the perceptions of the stake-holders into account. In the public perception, there is a bias towards risk reduction programs that reduce high consequence events while they tend to tolerate low consequence, high frequency events.

We examined a number of tools that can assist us in reducing the quantified risks. They help reduce the consequence or probability of events, sometimes both. Some are applicable in specific circumstances, while tools such as RCM have a wide range of applicability. Some tools such as HAZOP, TPM, RCA, and RCM are team activities. This fosters ownership and team spirit, which are important benefits that justify their higher costs. Some of these tools can help identify causes of human failures and are, therefore, very useful. Others are useful in the design phase, where the stress is on improving operational reliability. In the operational phase, RCM and TPM are appropriate, though we can also use the others selectively.

These tools help identify the applicable maintenance tasks and their timing. Thereafter, we have to go out and do it, in time and to acceptable quality standards. Only then will we reap the benefits of all this planning effort. Compliance is therefore very important and should be measured and reported regularly. An important role of the maintenance manager is to spot deviations in compliance and take corrective actions. Finally, if the work quality is poor, no amount of planning and compliance will help improve performance. It is essential to train, test, and motivate the workforce so that they reach acceptable quality standards.

We have noted the significance of perceptions and how they matter. Fear of the unknown and dread are two important factors that influence our perceptions. We can address the first concern by communicating our risk reduction plan to the stakeholders effectively. However, if we are not careful, it is easy to sound alarmist and this can raise the feeling of dread. There is an element of tight-rope walking when communicating risk management plans. Openness, a willingness to admit errors, and to have plans of action ready, all tend to build confidence. Good integration with the community, not merely with financial support, but also with active participation in their affairs helps build trust.

Our objective is to reduce risks to a tolerable level economically. In this chapter, we examined some of the tools at our disposal and their applicability. Tools alone do not suffice, and competent and motivated people must use them in the planning and execution of maintenance work.

References

1) Kletz, T. 1999. *HAZOP AND HAZAN*. Hemisphere Publishing, ISBN: 1560328584.
2) Davidson, J. 1994. *The Reliability of Mechanical Systems*. Mechanical Engineering Publication, Ltd. ISBN: 0852988818. 73-77.
3) Hoyland, A., and M. Rausand. 1994. *System Reliability Theory.* John Wiley and Sons, Inc. ISBN: 0471593974. 81-93.
4) Kepner, C.H. 1981.*The New Rational Manager.* Princeton Research Press, reprint edition. ISBN: 0936231017.

5) Cullen, W.G. 1991. The Public Inquiry into the Piper Alpha Disaster. Department of Energy, HMSO Publication Centre. ASIN: 010113102X.
6) Swain, A.D., and H.E. Guttmann. 1983. *Handbook of Human Reliability Analysis with Emphasis on Nuclear Power Plant Applications*. NUREG/CR-1278-F SAND80-200. 3-38.
7) Willmott, P. 1995. *Total Productive Maintenance The Western Way.* Butterworth Heinemann. ASIN: 0750619252.
8) Smith, A.M. 1992. *Reliability Centered Maintenance*. Butterworth-Heinemann. ASIN: 007059046X.
9) Edwards, A.W.F. 1992. *Likelihood*. Johns Hopkins University Press. ISBN: 0801844452
10) "Activists Found Guilty of Libel vs. McDonald's." Internet web site, reference http://www.pb.net/spc/mii/970742.html
11) "McJustice in Britain." 1997. *The National*. Internet web site, reference http://www.thenation.com/issue/970714/0714gutt.html. September 14.
12) "McLibel Suit Hits the Top of Civil Records in England. 1995. *Chicago Tribune*. Internet web site, reference http://www.mcspotlight.org/media/press/tribune_1feb97.html. December 11.
13) Tober, Bruce. 1995. "McDonalds Breaks Agreement in McLibel Suit." *Albion Monitor/News*. Internet web site, reference http://www.monitor/0-2-95/McLibel.html. September 2.
14) Nowlan, F.S., and H.F. Heap. 1978. *Reliability-Centered Maintenance*. U.S. Department of Defense. Unclassified, MDA 903-75-C-0349.

Further Reading

1) Anderson, R.T., and L. Neri. 1990. *Reliability Centered Maintenance: Management and Engineering*. Elsevier Applied Science Publishers, Ltd. ASIN: 185166470X.
2) Jones, R.B. 1995. *Risk-Based Management: A Reliability-Centered Approach*. Gulf Professional Publishing. ISBN: 0884157857.
3) Moubray, J. 2001. *Reliability-Centered Maintenance*. Industrial Press, Inc. ISBN: 0831131462.
4) *Risk, Analysis, Perception and Management*. 1992. The Royal Society. ISBN: 0854034676.
5) Nelms, C.R. 1996. *What You Can Learn From Things That Go Wrong*. C.Robert Nelms. ISBN: 1886118108.
6) Latino, R.J., and K. Latino. 2002. *Root Cause Analysis: Improving Performance for Bottom-Line Results*. CRC Press. ISBN: 084931318X.

CHAPTER 11

Information for Decision Making

The operating context of the business process will evolve and change throughout its life cycle. This is because external conditions are market driven and technological advances affect the business process. Fashion and changing customer preference influences the demand for products. Within the business, conditions may also change, with changes in ownership interests, new product lines, and occasionally, geographical relocation.

There are two objectives common to businesses, namely, to remain in business and to make a profit. In order to do that, businesses must be able to predict the market for their products. The greater this ability, the more successful they will be in adapting to the changing needs of the customers. While a feel for the market or instinct is a useful gift, it is only available to a few lucky entrepreneurs. The rest have to rely on their ability to gather the appropriate data and analyze it to obtain the required information. The lucky few also work hard at it, and one might argue that their success is due to this effort, though others may attribute it to their instincts.

Analysis by itself has no value. It must help achieve business objectives. For this purpose the data must be appropriate, analysis technique suitable, and the errors recognized and compensated. The resulting information is useful for making good decisions.

Time is a key element in any decision-making process. It places a limit on the pace of gathering and analyzing data. We have to make decisions even when the information is incomplete or not entirely accurate. With incomplete or incorrect information, there is a greater probability that we will make poor decisions. Time pressures are invariably high, so data quality and timeliness are always at a premium. The analyst must identify these risks when presenting recommendations.

11.1 WORK AND THE GENERATION OF DATA

Whenever we do any work, we obtain some data along with it, as listed below:

- Data about inputs, e.g., materials, labor, and energy consumption;
- Output volumes;
- Process speed data, e.g., start and finish times, cycle time;

- Process quality data, e.g., rejection levels, frequency of corrections, rework;
- Energy efficiency records;
- Process slowdowns, upsets or trips, direct and indirect delays;
- Data on soft issues, e.g., morale, attitudes, team spirit, customer satisfaction.

In addition, some relevant external data is being generated continuously by competitors, trade unions, customers, and government. It is better to analyze the two data sets separately, and use both sets of information in making decisions.

11.2 THE COLLECTION OF QUANTITATIVE DATA

Data may be numerical, in coded format, and in free text. Work history records are often in free text, but most other quantitative data is invariably in numerical or coded form. Process history is often in free text, but both work history and process data additionally contain a fair amount of numerical data.

The accountants and tax collectors were the first to recognize the importance of data collection. As a result, accountants designed data collection systems for their own use. These systems fulfill their original function, which is to record past performance and to ensure that an audit trail is available. The double-entry book-keeping system they designed was able to account for every cent. For this, they needed time, and some delays were acceptable in the interests of accuracy,

Most people are reluctant to design new data collection systems when there are existing systems in place. These are not always appropriate for their new decision-making roles, so they make attempts to bend the systems to suit. However the problem is more fundamental as the two have different functional requirements. The architecture for recording money transactions is not always suitable for analyzing failure history satisfactorily. In the latter case, the records must center on the equipment tag number. The equipment constructional details, operating context, performance, downtime, resources used, and cost data are all important, and must relate to the tag. We need the start-stop timing of events when we calculate equipment reliability parameters. Accounting systems do not usually demand these records so they are not always suitable for taking maintenance decisions.

From the maintenance engineers' point of view, a better approach would be to start by defining the function that they want performed. This top-down approach will help identify the type and timing of information required. They can then identify the data required for obtaining this information. By examining the existing data collection systems, they can check if they provide the required data at the right time. If so, there is no problem, otherwise they have to fill the gaps between desired function and that available with existing systems. If this is not possible, they have to design and install new systems.

Open architecture data bases can provide a solution that meets the requirements of both types of users. Systems that can talk to each other are superior to stand-alone systems. With suitable links, we can relate cost, history, equipment tag, and plant groups or other data collection nodes. This will help prevent the proliferation of systems and wasteful effort in recording the same data two or three times, along with the possibility of inconsistencies between systems.

Quantitative data for use in reliability calculations may be collected within one plant, several plants in one company, or as a joint industry project (JIP) by several companies. An example of such a JIP is the offshore oil and gas industry called OREDA[1], which has been very successful. The reliability data from OREDA is used, for example, in risk assessments, mathematical modeling, and RCM studies. Their data collection methodology has now been captured in an International Standard, ISO 14 224 1999.

11.3 THE COLLECTION OF MAINTENANCE DATA

In Chapter 4, we discussed failures at the component, equipment, sub-system, and system levels. We know that maintenance can restore performance to the design capability, but any enhancement beyond this level requires some redesign. There are two ways of enhancing equipment performance, first by reducing failure rates, and second by reducing the consequence of failures. While both methods are possible, each has an associated cost.This additional dimension means that there is a cost effective optimum solution awaiting discovery.

We can state these requirements as a set of functional requirements, as follows:

1. To identify design improvements to reduce failure rates;
2. To plan and execute maintenance in such a way that the consequences of failures are acceptably low;
3. To do the above at as low a long-term life-cycle cost as possible.

11.3.1 Failure reduction

This first function requires an analysis of all significant failures to establish their root causes. We need some or all of the following to analyze failures properly.

- Comprehensive and good quality incident investigation reports;
- Knowledge of the process; flow schemes, production rates, and other related data;

- Procedures used to operate the equipment, including start-up or shutdown sequences;
- Records of the actual operating history, including process charts and readings;
- History records showing failure and repair data;
- Spare parts consumption history;
- Information regarding the external environment, such as enclosures or weather conditions;
- Information about company culture, management style, worker attitudes and related soft issues;
- Knowledgeable resources to carry out the investigations.

Using root cause analysis (RCA), solutions follow fairly easily once we complete the study. The analysis must be thorough, and should not stop at proximate causes. It is easy to fall into this trap; often the RCA work stops at an early proximate cause. Eliminating proximate causes is like treating a sick person's symptoms instead of the disease itself. The analysts need patience and persistence to reach the underlying root causes.

The solutions may relate to the process, people, procedures, or plant. Often, the solution will involve training people, adjusting or revising procedures, or making the process steady. The solutions often require us to address management styles, company cultures, or conflicting goals. What do you think of an organization that proclaims 'Safety and Environment First' as its policy, and then punishes the supervisor or manager who decides to shut a plant down to prevent an event escalation? In hindsight, one may differ with the manager's judgment, but punitive action sends very strong messages to the entire workforce. Organizations that do not 'walk the talk' confirm the worst fears and doubts of their staff. In the Piper Alpha disaster, the offshore installation managers of Claymore and Tartan were aware of the mayday message from Piper, so they were aware of the major emergency there. However, they continued to produce at full capacity. Was safety or production higher on their agenda? Was there an underlying reason that could explain their actions?

Sometimes, as a result of the analysis, we may need to change the plant configuration or design. The implementation of these actions is itself a difficult issue. People resist change, even if it is in their own interest. Change management is a complex problem, and we must involve the workforce in the decision-making process itself, and in all stages of implementation.

11.3.2 Reducing the consequence of failures

We need a suitable set of maintenance strategies to minimize the consequence of credible failures. We can break this down into the following sub-functions:

1. To identify credible failure modes and their consequences;

2. To find applicable and effective strategies that can prevent or mitigate these consequences;
3. To create maintenance routines that integrate these strategies into practical and executable steps;
4. To measure and confirm that the routines are carried out to the required quality standards and at the right time.

As discussed in Chapter 10, we can use analysis tools such as RCM to achieve these objectives. What do we need to carry out these tasks? The data requirements include all those given in Section 11.3.1, as well as the following:

- Configuration of the equipment, e.g., series or parallel voting systems;
- Equipment performance data;
- Equipment layout drawings;
- Expected performance standards;
- Operating mode, e.g., duty/standby loading levels, continuous or intermittent operation;
- Knowledge of consequence of failures;
- An appropriate analysis tool.

Item 3 above requires us to match the maintenance routines with the strategies devised earlier. A competent maintenance planner equipped with suitable tools can do this work effectively. In order to check that the routines are in line with the strategies, we require an audit trail. The documents providing this trail constitute the relevant data.

Item 4 above requires us to measure the quality and timeliness of execution. We can achieve this if data about the following are available:

- Compliance records, to verify that the planned work is done in time;
- Staff training and test records to confirm competence;
- Service level records with respect to supporting logistics;
- The operating performance of equipment, as recorded after maintenance;
- Housekeeping and walk-about records, noting leaks and unsafe conditions;
- Results of physical audits carried out on maintenance work.

11.3.3 Cost data

We use systems built by the accountants, and they are experts on measuring costs. So it ought to be easy to measure maintenance costs. In practice the real maintenance costs are often quite difficult to obtain. The problem lies in defining the elements of cost that we should include under the heading *maintenance*. Distortions occur due to a variety of reasons, and a few examples will illustrate this point. Maintenance costs often include those related to the following types of work:

- Connection and disconnection of temporary equipment, such as mobile generators and provision of fuel and lubricants to such equipment;
- Simple low cost plant changes;
- Replacement of electrical motors (instead of repair);
- De-bottlenecking projects where existing equipment is used, but some components are modified or enlarged to increase the plant capacity;
- Spare parts that are withdrawn from stores but not used and often not returned for credit;
- Spare parts that are written off on receipt, even though they are not consumed;
- Operational tasks carried out by maintenance staff;
- Maintenance tasks carried out by operators;
- Accruals that do not reflect the real carry-over values.

There are fiscal incentives or tax breaks which encourage the creation of some of the distortions. In many cases, the value of each distortion may be relatively small. Taken as a whole, they could alter the cost picture, and because of inconsistencies from year to year, there may be apparent maintenance cost improvements without any real change in performance. Similarly, the books may show a worsening maintenance cost picture without any real change in performance. Cost management is always high on the agenda, and managers often think they are managing maintenance costs, without a full appreciation of some of these pitfalls.

A different type of distortion is possible in industries that have shutdown or turnaround cycles. We execute large volumes of maintenance work during these shutdowns, with the associated high costs. Thus, there are peaks and troughs in maintenance costs, but we enjoy the benefits over the whole of the shutdown cycle. Hence, a better way of treating such cyclical costs is to amortize them over the cycle length.This is usually difficult, since it means that we have to keep two sets of books, one for financial accounting, and the other for evaluating maintenance performance.

If we wish to control maintenance cost performance, the costs must be true and not distorted. The first step in this process is to define and measure the parameter correctly. We may need adjustments to compensate for shutdown cycles or inaccurate accruals. Transparency and consistency over the years are essential, if the figures are to be believable. Since you can only control what you can measure, it is important to measure the real costs directly attributable to the maintenance work.

Financial accounts must be accurate. This may require additional effort and time. Maintenance managers need a quick feedback of costs and commitments to do their jobs effectively. We can sacrifice some accuracy in order to obtain information quickly.

Our objective is to minimize the overall risks to the organization. If maintenance cost figures are unreliable or fudged, we expose ourselves to the risk of reducing essential maintenance when faced with pressures to contain costs. As a result, the risks of increased production losses and reduced technical integrity can rise.

11.4 THE COLLECTION OF QUALITATIVE DATA

In Chapter 7, we discussed the word qualitative in its descriptive sense. These relate to the factors that affect feelings and emotions of the people involved. They are responsible for morale and may help or hinder motivation. People do not always make decisions on sound rational judgment and analysis. Quantitative analysis can only go so far, and perceptions and emotions can easily swing the balance. This is why morale and motivation are important.

There are a few quantitative indicators of morale such as trends of sickness and absenteeism. Organizations experiencing high absenteeism among the workforce often find a similar trend among the supervisors and middle managers. This is often indicative of low morale. Other indicators include participation levels in suggestion schemes and voluntary community projects. A well recognized but hard to measure indicator is the number of happy faces around the facility. In an article entitled 'It's the manager, stupid,' *The Economist*[2] *reports on the results of a very large survey on employee satisfaction carried out by Gallup, the opinion-polling company. This covered over 100,000 employees in 24 large organizations over a 25-year period. They report that the best performing units were those where the employees were the happiest. The worst performers were also full of dissatisfied workers. The study also found that individual managers matter, by correlating employee satisfaction with things within their managers' control.*

Good morale is necessary for a motivated workforce. However, there are other factors as well, so it is not sufficient to have just high morale. These include the physical and psychological needs of people, as well as their domestic and social stability. Such factors are not easy to measure, since even the persons directly affected may not recognize them. These needs are also changing over time, and not in a linear or predictable way. You can recognize motivated people when you meet them. They are usually go-getters with a can-do attitude. They have ideas and are willing to share them. Often they are quite passionate about their ideas. Some of them sing or whistle at work. In spite of all these indicators, motivation is hard to measure, and we usually need expert professional help.

People with a logical frame of mind tend to shy away from such soft issues. Their zone of comfort is in rational thinking, preferably with numbers to support their decisions. Their contribution is in countering those who decide by hunch and gut-feel.

Morale and motivation are hard to measure, and the results may make us feel uneasy. These are some of the reasons why we do not always address them satisfactorily. The point however is this, if you do not know what makes people tick, you are not always able to make the right decisions.

We should monitor sickness and absenteeism regularly. These records are easy to collect and are useful in judging morale. We should measure motivation periodically with the help of professional experts. The trends will help decide if we need corrective action.

11.5 ERRORS IN DATA COLLECTION

The quality of any analysis is dependent on the correctness of the source data. However good the analysis technique, if serious errors exist in the raw data, the results will not be of much use.

We can categorize maintenance records into two main types:

1. Static data, including tag numbers (which identify the items of equipment by location), make, model and type descriptors, service details, and cost codes;
2. Dynamic data, including vibration levels, operating performance, time of stoppage and restart, as-found condition, repair history, spare parts, and resources used.

Errors in static data are usually reconcilable as it is possible to spot them through audits. If the tag number entry is incorrect, for example, if pump P4120A is recorded as P4210A, we can use the service or duty to validate it. If on the other hand, we record P4120A as P4120B, we can use the operating log to reconcile this error. Similarly, we can identify an error in the cost code by identifying the tag number and hence the location and service. The relative ease with which we can verify static data makes them less critical, as long as a logical numbering system has been used. This does not reduce the need to record static data correctly in the first instance. If the error rate is high, the validation task can become very difficult.

Dynamic data is more difficult to validate or reconcile. Some dynamic data such as vibration or alignment readings are volatile. You cannot come back a few days or weeks later and obtain the same results, since they will have changed. In other cases, the record exists only in one place. For example, the technician records the as-found condition or repair history only in the job card. Similarly, if there is some confusion between the active repair time and the downtime, it may be impossible to validate. Some dynamic data entries are duplicated. In these cases, one can trace the errors easily. For example, spare part consumption details may also be available in warehouse or purchase records.

Human eyes can easily pick up text data errors. These include errors such as spelling mistakes, keystroke errors, transposition of letters or words, use of hyphens, backslashes, or colons between words. If we use conventional software to search for such errors, the task is very difficult. Such software cannot handle word order, differentiating between, for example, blind venetian and venetian blind. However specialized pattern-recognition software is now available. Desktop computers are powerful enough to use them effectively. The software has built in rules of forgiveness, and a lexicon of words with similar meanings that it can use to expand the searches. Other features include context sensitivity, and the ability to use conditional logic (...if...so....), and change of endings (...ing, ...en ...er, etc.), without the need for wild card searches. As a result, the search quality approaches that of the human eye, but is obviously a lot faster. With current technology, we can manage errors in text data entry effectively.

One can code data at source. This may consist of two-to-ten letters or numbers that represent a block of data. The main data fields are as follows;

- Defect reported, e.g., running hot, stuck open, high vibration, spurious alarm or trip, external (or internal) leak, fail to start (or stop, open, close);
- As found condition, e.g., worn, corroded, broken, bent, dirty, plugged, jammed;
- Probable cause, e.g., process condition (pH, flow, temperature, pressure, plant upsets, foaming), procedures not followed, wrong installation, drift, misalignment, loss of calibration, quality of utilities;
- Repair description, e.g., part(s) replaced, cleaned, realigned, recalibrated, surface finish corrected, lubricated, resealed;
- Origin and destination of equipment;
- Technician's identification reference number or code name.

Coded entries are easy to analyze using simple spreadsheets.They are popular and are suitable for a range of applications. When used correctly, we can minimize errors and obtain results quickly.

11.6 FIXED FORMAT DATA COLLECTION

Many people see the use of coded entries or fixed format reporting as a solution to the elimination of errors in data collection. There are many advantages in using fixed formats. Some of these are:

- There is standardization in data collection, and its quality is less dependent on the competence or personal knowledge of the person collecting the data.

- There is a check-list or prompt available to guide the person;
- The time required to fill in a form or report is minimal;
- The time required to collate and analyze the data is minimal;
- It is easy to verify the completeness of the entries in the form;
- It facilitates electronic recording and analysis of data;
- It enables quick searches and simple statistical calculations.

As a result, there is a strong move towards the creation and use of fixed format reporting. A number of modern maintenance management systems use fixed formats, quoting the many advantages discussed above. Appendix 11-1 shows a table of codes that we can use in modern maintenance management systems. There are four main categories that we use to describe the failure details, as follows:

Series 1000	Failures as reported
Series 2000	Main work done
Series 3000-8000	Failures by equipment type
Series 9000	As found condition, fault found

The technician or operator must fill in all four categories in the appropriate columns. Multiple entries may be required in each of the categories to allow for the different scenarios. These entries only relate to the failure details. In addition, the form will have dates, account codes, free text history, and other items discussed earlier.

There are however a number of drawbacks with fixed format reporting, as listed below.

- Data entry errors can easily occur due to the selection of the wrong code. The use of a wrong keystroke, or the selection of the wrong code number can occur easily, and seriously distort the information recorded;
- If the entries are by hand, the person reading it later may misunderstand the hand-writing;
- It is quite common to provide a pick-list to help the technician in entering the data. Providing only two or three alternatives is usually not adequate. The choices tend to grow, and the pick-lists often contain six or more items from which to make the selection. As a result of boredom or disinterest, the recorder may choose the first or second item in the pick-list each time. Such behavior defeats the purpose of providing multiple choices;
- It is difficult to describe some entries even with six or more choices. The available options can never fully describe every event or observation. In such cases, an item called *other* or *general* is justifiable. When such an option is available, it is common to find many entries falling in this cate-

gory. This becomes a catch-all or sink-code into which the majority of entries fall.

The main problem with fixed formats is that it is not possible to identify source data-entry errors. Earlier, when free text was time consuming and laborious to analyze, use of fixed-formats was justifiable. The speed and accuracy of analyzing free text with the software tools currently available makes fixed format reporting less attractive. The quality problems associated with them need to be recognized and resolved.

11.7 OBTAINING INFORMATION FROM DATA

In the context of maintenance management, the information we require relates to one of the following areas:

- Output of maintenance work, namely, system effectiveness, plant availability, reliability and efficiency;
- Inputs such as labor hours, materials, and energy;
- Information to improve intrinsic reliability by, e.g., identifying the root causes of failures;
- Information to demonstrate timely completion of maintenance work;
- Information to assist in the planning of maintenance work in future.

In each case, we have to analyze the appropriate set of data suitably. We will consider each of them in turn.

- We measure system effectiveness in volumetric terms, namely, how much we produce versus how much we require and what it is possible to produce. Usually we can apply this metric at the plant level or at system level, but is difficult to apply it at the equipment tag level. Because of this difficulty, we use the time-availability or the proportion of time the equipment is able to produce, to the total period in operation. The latter metric requires the start and end dates, and the duration of downtime for planned and unplanned maintenance work. If a good maintenance management system is in place and the records are available, this data is easy to obtain. Otherwise we may need to trawl through the operating log and the maintenance supervisor's note book.
- A simple metric to use to judge the plant and equipment reliability is the mean time to failures or MTTFs. To do this, we simply divide the time in operation by the number of failure events. Often, the time in operation is not always available. So we make a further simplification and use calendar time instead. At the plant or system level, we can measure the number of trips and unplanned shutdowns. The time in operation will be the calendar time less the duration of any planned shutdowns. While the absolute val-

ues are of interest, trends are even more important. A rising trend in MTTFs is a sure indication of the success of the improvement program. Sometimes, even these measurements are not possible, but maintenance work orders (or job cards) may be available. We can calculate the mean time between non-routine work orders as a measure of reliability. Here 'non-routine' means work orders for corrective and breakdown maintenance work. Each of these approximations decreases the quality of the metric. However, in the absence of other data, these may be the best available.

- The operators will normally monitor plant efficiency continuously. The metrics include flows, energy consumption, pressure or temperature drops, conversion efficiency, and consumption of chemicals and utilities. Efficiency is one of the parameters where the deterioration in performance shows an incipiency curve that operators can plot quite easily. Since the loss of efficiency is a strong justification for a planned shutdown, it is a good practice to monitor this parameter.
- Records of inputs such as human resources, energy, and materials are normally available. It should be possible to identify the inputs at the equipment, system, and plant levels.
- It is a good practice to record all near-misses and incidents. We need these to carry out root cause analysis. We should analyze high-risk potential operational and integrity-related events. Since the RCA work may start several weeks after an event, the quality of incident reports is important.
- Technicians should record the start and completion of preventive and corrective maintenance work in the maintenance management system. We define compliance as the ratio of completed planned work to that originally scheduled. The monitoring of compliance is important, and can normally be produced with data from the maintenance management system;
- Learning is a continuous process. On each occasion that we do work, new learning points arise. If we capture and incorporate the learning points in the next plan, we complete the continuous improvement loop. A mechanism for capturing these learning points is therefore necessary. We can use the maintenance management system itself for this purpose or build a separate database.

11.8 DECISION SUPPORT

We have to manage the planning and execution of maintenance work properly. Maintenance professionals must recognize the importance of data in the continuous improvement process. Improvements in maintenance performance depend on course corrections based on proper analysis of data.

11.9 PROCEDURES

In chapter 8, we discussed the role of procedures in preventing the escalation of events. They enable the transfer of knowledge and serve as training material for staff at the time they need them. The best results are obtained when they are easy to understand, accessible to the people who need them and are updated regularly. For example, when startup and shutdown of critical equipment is difficult, it is useful to have these procedures in weatherproof envelopes at site. Operators and maintainers should be able to read the procedures they need in their supervisors' offices. In high performance organizations, one is more likely to find well-thumbed copies of procedures. Pristine copies of procedures are a cause for concern, not a matter of pride.

Keeping procedures up-to-date takes effort, discipline and resources. Revisions may be triggered by undesirable incidents or advice from equipment vendors. All procedures should be vetted periodically on a revision schedule, and revisions must be dated. This activity is important enough to be mentioned in the job description of the maintenance manager.

11.10 BUSINESS PROCESS MANAGEMENT

In some organizations, information requirements may be poorly defined or resourced. Sometimes this is due to a poorly defined business process. Such situations can lead to poor management control. If any of these deficiencies are identified, some of the tools we discussed earlier, for example, IDEF, RCA, FTA, or FMEA can be used to rectify the situation.

11.11 CHAPTER SUMMARY

In order to stay in business, managers must be able to adapt to changing circumstances. Some may be in their control, but market forces can affect the operating context with the passage of time. Often we have to make decisions with limited information.

Whenever we execute work we generate some useful data. We gather and analyze this data to make the appropriate decisions. During this process, we may introduce a variety of errors. Some people believe that fixed format reporting will solve many of these errors. We noted the flaws in this argument and the benefits and drawbacks of fixed format reports. Free text is now quite easy to read with software and offers an alternative method.

In order to manage maintenance effectively, we need information in the following key areas.

- Outputs from maintenance, e.g., plant availability and reliability.

- Inputs to maintenance, e.g., materials, resources and costs.
- Compliance with preventive maintenance program.
- Information to help improve operational reliability, e.g., data for computing MTBFs, MTTRs, RCM and RCA studies.
- Information to improve productivity, e.g., delays, rework and their causes.
- Information to improve planning of future maintenance, e.g., as-found condition on opening equipment, success rates on tests etc.

Maintenance data collection and analysis are important in supporting decision making.

Reference

1) OREDA. Internet web site, reference http:lloreda.com
2) "It's the Manager, Stupid." 1998. *The Economist*. August 8: 68.

Appendix 11-1

FIXED FORM DATA—CODES AND DESCRIPTIONS

Failures as reported

Operational failures

1001	Fail to start
1002	Stopped while running/Trip
1003	Low output
1004	Operating outside design
1005	Poor startup procedure
1006	Poor shutdown procedure
1007	Stuck open/close
1008	External leak
1009	Internal leak

Mechanical failures

1011	Worn
1012	Leakage
1013	Vibration/noise
1014	Blocked/fouled
1015	Stuck open/close
1016	Overheated/burnt
1017	Impact

Material failures

1021	Corrosion/erosion
1022	Fatigue
1023	Fracture
1024	Ductile/plastic deformation
1025	Incorrect materials

Electrical failures

1031	No power/voltage
1032	Earth fault
1033	Short circuit
1034	Open circuit
1035	Burnt
1036	Contacts welded

Instrument failures

1041	Out of adjustment
1042	Leakage
1043	Control failure
1044	No signal/indication/alarm
1045	Faulty signal/indication/alarm
1046	Common mode failure

Design related causes

1051	Not operator friendly
1052	Not per standards
1053	Operation outside design
1054	Not fail -safe

External causes

1061	External environment
1062	Blockage/plugged
1063	Contamination
1064	Upstream/dowstream equipment
1065	Unprotected surface

Miscellaneous causes

1071	Unknown cause
1072	Combined causes
1073	New cause-describe

Main work done

2010	Replace
2020	Restore/repair
2030	Adjust/align/calibrate
2040	Modify/retrofit
2050	Check/inspect/monitor condition
2060	Combination of repair activities

Failures by equipment type

Pump unit - centrifugal, rotary

3011 Rotor assembly
3012 Casing
3013 Impeller/rotor
3014 Bearing
3015 Coupling
3016 Shaft
3017 Shaft mechanical-seal
3018 Balancing drum
3019 Wear rings, bushes
3020 Other items - specify

Pump unit - reciprocating

3021 Piston, piston rings
3022 Suction/delivery valves
3023 Cylinder, casing, liner
3024 Bearings
3025 Shaft seals
3026 Diaphragm
3027 Auxiliaries
3028 Control System
3029 Lubricator
3030 Other items - specify

Compressor unit - centrifugal

3031 Rotor assembly
3032 Casing, barrel
3033 Impellers
3034 Bearings
3035 Coupling
3036 Shaft
3037 Shaft mechanical seal
3038 Lubrication system
3039 Seal oil system
3040 Control systems
3041 Other items - specify

Compressor unit - reciprocating

3051 Piston, piston rings, vanes
3052 Suction/delivery valves
3053 Suction unloader
3054 Cylinder, casing, liner
3055 Bearings
3056 Shaft seals
3057 Diaphragm
3058 Auxiliaries
3059 Lubricator
3060 Control System
3061 Other items - specify

Gas Turbines

3071 Burners, combustors
3072 Transition piece
3073 Fuel gas supply
3074 Fuel oil supply
3075 Air compressor
3076 Gas generator
3077 Power turbine
3078 Blades
3079 Bearing
3080 Coupling
3081 Gear box
3082 Air filter
3083 Lubrication system
3084 Starting unit
3085 Casing
3086 Fire protection system
3087 Ventilation fan
3088 Acoustic hood
3089 Turbine control system
3090 Other items - specify

Steam Turbines

3091 Trip and throttle valve
3092 Steam chest valve
3093 Governor
3094 Casing, barrel
3095 Rotor
3096 Blade
3097 Bearing - radial
3098 Bearing - thrust
3099 Hydraulic system
3100 Coupling
3101 Gear box
3102 Lubrication system
3103 Shaft seal
3104 Condenser
3105 Vacuum pump
3106 Other items - specify

Electrical Generator

3111 Rotor
3112 Stator
3113 Bearing - radial
3114 Bearing - thrust
3115 Exciter
3116 Cooling system
3117 Air filter
3118 Lubrication system
3119 Protective system
3120 Other items - specify

Electric Motor

3121 Rotor
3122 Stator
3123 Bearing
3124 Fan
3125 Ex-protection
3126 Starter
3127 Local push button station

3128 Control system
3129 Lubrication system

Internal Combustion Engines

3131 Air filter
3132 Fuel filter
3133 Fuel pump
3134 Injector
3135 Spark plug
3136 Starter
3137 Valve
3138 Manifold
3139 Piston, piston-ring
3140 Battery
3141 Radiator
3142 Water pump
3143 Control system
3144 Other items - specify

Starting system

3151 Electric motor
3152 Hydraulic motor
3153 Hydraulic pump
3154 Gear train
3155 Clutch
3156 Start control system

Columns, Vessel

3161 Pressure vessel
3162 Internals (trays, demisters, baffles)
3163 Instruments
3164 Piping, valves
3165 Nozzles, manways
3166 External appurtenances, access

Reactors, Molecular sieves

3171 Pressure vessel
3172 Internals (trays, catalyst/ ceramic beds)
3173 Instruments
3174 Piping, valves
3175 Nozzles, manways
3176 External appurtenances, access

Heating, Ventilation, Air Conditioning

3181 Fan
3182 Fire damper
3183 Filter
3184 Dryer/conditioner
3185 Gas detection system
3186 Control and monitoring system
3187 Refrigeration compressor
3188 Coolers, radiators, heat exchangers
3189 Motor
3190 Gear box
3191 Other items - specify

Power transmission

3196 Gearbox
3197 Coupling
3198 Clutch/Variable Drive

Boilers, Fired heaters

3201 Pressure parts
3202 Boiler/furnace tubing
3203 Burners
3204 Fuel system
3205 Electrical heating elements
3206 Insulation, Refractory lining
3207 Auxiliaries(air/water supply etc)
3208 Control and protective systems
3209 Valves
3210 External appurtenances
3211 Other items - specify

Heat Exchangers

3221 Pressure parts, process media
3222 Pressure parts, cooling medium
3223 Valves
3224 Electrical heating elements
3225 Auxiliaries
3226 Control and protective systems
3227 Other items - specify

Piping systems

3231 Pipe
3232 Flanges, fittings
3233 Instruments (Orifice plates, gauges)
3234 Insulation, paintwork
3235 Structural supports
3236 Other items - specify

Hydrocyclones

3241 Pressure parts
3242 Internals
3243 Nozzles, valves
3244 Control and monitoring system
3245 Other items - specify

Lubrication system

3251 Pump with motor
3252 Cooler
3253 Filter
3254 Valves and piping
3255 Reservoir
3256 Instrumentation/Accumulator
3257 Oil
3258 Other items - specify

Instruments-sensors

4001 Pressure
4002 Flow
4003 Temperature
4004 Level
4005 Speed
4006 Density
4007 Humidity
4008 Turbidity
4009 Proximity
4010 Other items - specify

Instruments - signal transmission

4011 Transmitters
4012 Receiver
4013 Integrators
4014 Junction boxes, marshalling racks
4015 Signal convertors
4016 Cables and terminations
4017 Tubing and connectors
4018 Other items - specify

Processing units

4021 Computers
4022 Amplifiers, pre-amplifiers
4023 Central processing units
4024 Analysers
4025 Computing relays
4026 Printed circuit cards
4027 Other items - specify

Display units

4031 Gauges - pressure, level, flow
4032 Alarm annunciators
4033 Klaxons, hooters
4034 Recorders
4035 Video displays
4036 Printers
4037 Other items - specify

Executive elements

4041 Pneumatic/hydraulic actuators
4042 Electrical actuators
4043 Valve positioners
4044 Control valves
4045 Trip and release mechanisms
4046 Other items - specify

Other Instruments

4051 Meteorological instruments
4052 Test equipment - pneumatic/ hydraulic
4053 Test equipment - electrical
4054 Other items - specify

Electrical distribution

5001 Transformers, Power factor capacitors
5002 HV circuit breakers
5003 LV circuit breakers
5004 Miniature circuit breakers, fuses, isolators
5005 HV switchgear
5006 LV switchgear
5007 Switchboards, cubicles
5008 Motor starters
5009 Junction and marshalling boxes
5010 Relays, coils, protective devices
5011 Other items - specify

Electrical heaters

5021 Process heaters
5022 Trace heaters
5023 Trace heater controls, switchgear
5024 Other items - specify

Electrical - general items

5031 Cables, jointing
5032 Cable termination
5033 Batteries
5034 Battery chargers
5035 Electrical test equipment
5036 Electric hoists
5037 UPS systems
5038 Rectifiers, invertors
5039 Cathodic protection systems
5040 Miscellaneous electrical items

Lighting systems

5051 Fluorescent fittings, bulbs
5052 Flood light fittings, bulbs
5053 Sodium vapor fittings, bulbs
5054 Mercury vapor fittings, bulbs
5055 Beacons fittings, bulbs
5056 Other items - specify

Miscellaneous process equipment

5061	Silencer
5062	Ejector
5063	Flare
5064	Hot oil system
5065	Tank, silo
5066	Runway beam
5067	Crane
5068	Chain block
5069	Slings, wire rope
5070	Other lifting equipment
5071	Conveyor
5072	Other items - specify

As found condition, fault found

9005	Worn
9010	Broken, bent
9015	Corroded
9020	Eroded
9025	Fouled, blocked
9030	Overheated, burnt
9035	Fatigued
9040	Intermittent fault
9045	Worked loose
9050	Drift high/low
9055	Out of span
9060	RPM hunting
9065	Low/high output voltage/frequency
9070	Short/open circuit
9075	Spurious operation (false alarm)
9080	Signal transmission fault
9085	Electrical/Hydraulic power failure
9090	Injection failure
9099	Other (specify in text field)

CHAPTER 12

Improving System Effectiveness

In Chapter 8, we looked at a number of case histories relating to major disasters. The sequence of events leading to these disasters appears to follow a common pattern. Production losses are also due to similar failures. When people ignore warning signs, process deviations or equipment failures may lead to loss of process control. If we do not resolve these in time, they can escalate into serious failures. We can reduce the risks of safety or environmental incidents and minimize production losses by improving the effectiveness of the relevant systems.

In order to reduce risk to tolerable levels, we need data and tools to analyze performance. With these elements in place, we can put together a plan to improve system effectiveness. In this concluding chapter, we will examine implementation issues, and see what practical steps we can take. In Appendix 12-1, we will look at applications outside the equipment maintenance area, using the more holistic definition of maintenance.

12.1 SYSTEM EFFECTIVENESS

When there are no constraints at the input and output ends of a plant, it can produce to its design capacity. The only constraint is its own operational reliability. The ratio of the actual production to its rated capacity is its system effectiveness. It takes into account losses due to trips, planned and unplanned shutdowns, and slowdowns attributable to the process or equipment failures. A simple way to picture this concept is to think of the plant or system being connected to an infinite supply source and an infinite sink. In this case, the only limitation to achieving design capacity is the operational failures attributable to the process, people, and the equipment. Thus if the process fluids cause rapid fouling, equipment fails often, or we do not have well-trained and motivated operators and maintainers, the system effectiveness will be low.

Limitations in getting raw materials, power, or other inputs, or poor market demand will influence our expectations from the plant. The scheduled production volume has then to be lower than the design capacity to allow for these limitations. If the actual production volume is lower than these reduced expectations, the system effectiveness falls below 100%.

Certain processes require the delivery of products on a daily, weekly or seasonal basis. For example, supermarkets have to bring fresh stocks of milk and vegetables daily. Similarly, a newspaper can receive incoming stories up to a given deadline. After this time, the presses have to roll, so new stories have to wait for the next edition. If an organization settles its payroll on a weekly basis, their accountant has to manage the cash flow to suit this pattern.

If there are delays in delivering the goods or services and if we do not meet the stipulated deadlines, we may incur severe penalties. In these cases, we can think of the demand as discrete packages or contracts. Time is of essence in these contracts, and is a condition for success. Delivery of the product beyond the deadline is a breach of contract. In this situation, we define system effectiveness as the ratio of the number of contracts delivered to those scheduled in a day, week, month, or other time period.

We defined availability (refer to Section 3.7) as the ratio of the time an item is able to perform its function to the time it is in service. A subsystem or item of equipment may be able to operate at a lower capacity than design if some component part fails. Alternatively, such failures may result in some loss of product quality that we can rectify later. In these cases, the operator may decide to keep the system running till a suitable time window is available to rectify the fault. The system will then operate in a degraded mode till we correct the fault. If the functional requirements were that the item produces at 100% capacity throughout, the degradation in quality or reduction in product volume means that the system no longer fulfills its function. Technically, it has failed and is therefore not available. In practice, however, the functional statement can be quite vague, so it is customary to treat an item as being available as long as it is able to run. Many people use this interpretation because of the lack of clarity in definition. With this interpretation, high availability does not always mean high system effectiveness. A second condition to be met is that there are no degraded failures that can bring down product volumes or quality.

The picture changes when we have to deliver discrete quantities in specified time periods. Here the timing of the degraded failure is important. If a fresh-milk supplier's packaging machine fails at the beginning of a shift, it may be possible to meet the production quota. For this, we have to do the repairs quickly and must have the ability to boost the production thereafter. If the same failure takes place towards the end of the shift, it may be impossible to fulfill the contract, even with quick repairs. This is because it takes some time to start and bring the process itself to a steady state. Assuming we can boost production, we still need some period of time to make up for lost volumes. As you can see, the timing of the failure is an additional parameter that we have to take into account. In order to raise the system effectiveness, we have to improve the operational reliability of the equipment and subsystems. This will reduce the total number of failures, so the frequency of production interruptions falls. When this is sufficiently low, we do not have to worry whether these take place at the beginning or end of the shift.

12.2 INTEGRITY AND SYSTEM EFFECTIVENESS

In Section 4.1.4, we discussed hidden failures. Protective equipment such as smoke or gas detectors, pressure relief valves, and overspeed trip devices fall in this category. They alert the operators to potentially unsafe situations. They may also initiate corrective actions without operator intervention. If these fail, a second line of defense is available to limit the damage (refer to Section 9.2). By maintaining these safety systems properly, we can reduce the chances of event escalation. The barrier availability is a measure of the effectiveness of safety systems. In terms of the risk limitation model, a high barrier availability helps us achieve integrity.

12.3 MANAGING HAZARDS

12.3.1 Identification of hazards

The hazards facing an organization may relate to its location, the nature of materials it processes or transports, and the kind of work it executes. In addition, there may be structural integrity issues related to the equipment used. Process parameters (pressure, temperature, flow, speed, toxicity, or chemical re-activity) can influence the severity of structural hazards.

Once we have identified these hazards, we have to assess the level of risk involved. These risks may be qualitative or quantitative, and we must assess them using appropriate methods.

12.3.2 Control of hazards

If we have a method to reduce the process demand, we should do this in the first instance. In terms of the event escalation model, we try to reduce the rate of occurrence of minor failures or process deviations. Techniques such as HAZOP or root cause analysis are useful in reducing the probability of occurrence of process deviations.

If this is not possible, we try to improve the availability of the *people, procedure,* and *plant* barriers. As we have seen in Chapter 9, this helps reduce the escalation of minor events into serious ones. Note that we should improve those elements of the barrier that are most effective. For example, if there is a hot process pipe that could cause injury, consider providing a *plant* barrier such as insulation instead of a *procedure* such as a warning sign.

12.3.3 Minimization of severity of incidents

If a serious event has already taken place, we have to try to limit the damage. We used the damage limitation model to explain why we need a high availability of the *people, procedure,* and *plant* barriers at this level.

In trying to manage barrier availability, we can work with one or both of the variables, namely the intrinsic reliability and the test frequency. The age-exploration method we discussed in Section 3.10 is of use when reliability data is not readily available. There is a tendency for companies to introduce additional, often illogical, maintenance checks after an incident in an attempt to prove they have done something. From the discussion in chapters 8 and 9, we know that they have to invest in improving the availability of the event escalation barriers. The maintenance manager is in a good position to lead the way.

12.4 REDUCING RISKS - SOME PRACTICAL STEPS

12.4.1 Appreciating life cycle risks

Awareness of the risks we face is merely the first step. The key players—namely, senior management, staff, union officials, pressure groups, and the local community—also have to agree that these are risks worth addressing. Issues that affect safety and the environment are relatively easy to communicate and the escalation models can assist us in building up our case. Improving plant safety, reliability, and profitability will appeal to all the stake-holders as worthwhile objectives. Since people favor risk reduction programs that reduce high consequence events, we must align our efforts accordingly. In communicating our risk reduction program to the community and to the workforce, we have to address two important factors tactfully. These are fear of the unknown and dread, as discussed in Section 7.3. The information must be truthful and reduce the fear of the unknown, without raising the sense of dread.

12.4.2 Tools and techniques

In Chapter 10, we examined a number of tools that can assist us in reducing the quantified risks. These include, for example, HAZOP, TPM, RCA, and RCM. Some of these tools can also help identify causes of human failures and are, therefore, very useful.They may reduce the consequence or probability of events, sometimes both.

Some tools are useful in the design phase where the stress is on improving operational reliability. Other tools are applicable for use in the operational phase. These help us plan our maintenance work properly so that we avoid or mitigate the consequences of failures.

12.4.3 The process of carrying out maintenance

In Section 9.3, we discussed the continuous improvement cycle and its constituent maintenance phases. These are planning, scheduling, execution, analysis, and improvement. In order to achieve high standards of safety, we have to plan and schedule work properly. We use toolbox talks and the permit to work system to communicate the hazards and the precautions to take. Protective safety gear and apparel will help minimize injury in the event of an accident. The quality of the work has to be satisfactory. Quality depends on knowledge, skills, pride in work, and good team spirit of the workforce. Staff competence, training, and motivation play an important role in achieving quality work.

12.4.4 Managing maintenance costs

In managing maintenance costs, we noted that the main drivers are the operational reliability of the equipment and the productivity of the workforce. If we manage these drivers effectively, the costs will fall. However the workforce, unions, and community may view maintenance cost reductions with suspicion. Hence, these cost savings must be a natural outcome of the reduction in failure rates. This reduces the volume of work, and better methods improve the productivity. The combination will help reduce costs, while improving the equipment availability. By demonstrating that the cost savings are a natural outcome of these actions, we can allay the fears of the interested parties.

The most effective approach is to reduce failure rates first and then tackle productivity issues. In practice we find a reversal in emphasis and often the focus is on productivity aspects. In our eagerness to improve productivity, we might eliminate some essential maintenance, thereby increasing the risks to the organization. This is also the public perception and one of the reasons for resistance to risk reduction programs.

One can improve productivity by good quality planning and scheduling. Using industrial engineering tools such as method-study, one can improve planning and scheduling so as to enhance productivity. Managers can contribute to reducing the idle time of workers as they control planning and scheduling resources. Unfortunately, people often use time-and-motion study in preference to method-study. As a result, instead of eliminating unnecessary activities or transport, they only try to speed up the work. A sweat-shop mentality will not be effective or find favor with the work force.

12.5 COMMUNICATING RISK REDUCTION PLANS

Good intentions do not necessarily produce good results. The actions we propose may not appeal to the target group. When we explain to people how they can reduce their personal risks, they do not necessarily follow the advice. Recall our earlier comment that people tend to believe that bad outcomes will affect others, but not themselves.

A transparent organization that is willing to share good news along with the bad news is more likely to succeed in communicating its position. Similarly, one that takes active part in the community is more likely to receive public sympathy if things go wrong. On the other hand, if it tries to soften the impact using professional spin-doctors, the public can become suspicious. We must tell the people who have a right to know. Tact is important, so that the message does not convey a sense of dread.

12.6 BRIDGING THE CHASM BETWEEN THEORY AND PRACTICE

There are many learned papers that address the application of reliability engineering theory to maintenance strategy decisions. Many of them use advanced mathematics to fine tune maintenance strategies. The authors have limited access to field data, and their recommendations are often abstract and difficult to apply. So these remain learned papers, which practitioners do not understand or cannot apply to real-life situations.

Maintainers are under pressure to improve operating performance. They have access to field data, but are often not aware of the tools and techniques that they need and which reliability engineers can provide. In many cases, they do not apply even basic theory, partly due to lack of familiarity and partly because the mathematics is beyond them. Similarly, designers should be able to select the optimum design option by applying, for example, reliability modeling. They may be unaware of the existence of these techniques or not have access to them.

This chasm between the designers and maintainers on the one hand and the reliability engineers on the other is what we have to bridge. Reliability engineers have to understand and speak the language of the maintenance and design engineers. They have to market the application of their knowledge to suit the requirements of their customers. For this purpose, they may have to forsake some of their elegant mathematical finesse. Ideally, if their models and formulae were user-friendly, the designer and maintainer could happily apply the techniques. Once they start applying these techniques successfully, there will be a feedback to the theoreticians, so both parties will benefit.

12.7 MAINTENANCE AS AN INVESTMENT

Maintenance is much more than finding or fixing faults. It is an essential activity to preserve and improve our technical integrity and to maximize our profits. In this sense, it is an ongoing investment that will bring in prosperity. We have seen how to approach it in a structured and logical manner, using simple reliability engineering concepts. In making their decisions, maintainers need timely cost information. Even if there are minor errors in this information, the decisions are not likely to be different. Maintainers aim to reduce the risks to integrity and production capability, and we know they can do so by improving the availability of the event escalation barriers.

Throughout history, people have tried to make perpetual motion machines and failed. Similarly, there are no maintenance-free machines. Investors who expect a life-long cash cow merely because they have built technologically advanced plants are in for a surprise. Even these need maintenance, which is an investment to preserve the health and vitality of their plant. There is a proper level of maintenance effort that will reduce the risks to the integrity and profitability of the plant to an acceptable value. We can optimize the costs related to this effort by proper planning, scheduling, and execution. Any effort to reduce this cost further will result in an increase in the risks to the organizational.

12.8 CHAPTER SUMMARY

In this final chapter, we examined system effectiveness and reviewed several examples. From an integrity point of view, system effectiveness of protective systems is very important. These are the Plant Barriers in our risk limitation model (Figure 9.1), and play a vital role in limiting event escalation.

This leads us to a discussion on the practical steps we can take to reduce risks; understanding, applying the right tools and techniques, and executing the maintenance work cost-effectively.

Doing the work is, by itself, not enough; we need to communicate our risk reduction strategies effectively to the concerned people in simple language. This affects perceptions, and it is as important an issue to manage as quantitative risk. We conclude this chapter emphasizing that maintenance is an investment, essential for managing the risks facing any organization.

12.9 BOOK SUMMARY

We started off this journey by asking **why** we do maintenance, so that a clear justification could be offered. Using the event escalation and maintenance models, we concluded that its raison d'être was to

i. preserve integrity and hence the long term viability of the plant
ii. ensure profitability by providing availability of the plant at the required level.

This approach should help maintenance practitioners prove the value of their work and justify maintenance cost as an investment towards viability and profitability.

Managers manage risk, which in its quantitative sense depends on two factors. We can learn about the **probability** of adverse events by knowing some basic reliability engineering theory. Therefore, we developed this theory using (mainly) tables and charts. We can do root cause analysis to reduce the probability of failure. Doing the right maintenance at the right time also reduces the probability of failure. In the main, however, when we do maintenance, we minimize the **consequences** of failure. We developed an understanding of the RCM process as we progressed through the book. With an RCM approach, we can say **what** maintenance tasks we should do, and **when** we should do them.

The qualitative aspects of risk, those dealing with **perceptions**, are also very important as they affect the way decisions are made. We explained why seemingly illogical decisions are made, based on the perception of the people involved. Rather than fight this behavior, we are better off adapting our own strategies, so that they appeal to the decision-makers.

Many of us struggle with cuts in maintenance budgets. Maintenance costs are always under pressure, so it is best if we addressed the cost drivers. These are the **operational reliability** of the equipment and the **productivity** of the maintenance staff. Operational reliability depends on both operators and maintainers. Operational philosophy, such as duty-standby operations can bring large improvements in reliability and costs. There are examples in the book of other operational philosophies, such as those relating to PRVs, which can be applied to advantage. Good maintenance work contributes greatly to operational reliability. For this we need strategies, knowing **what** work is worth doing and **when** we should do them. We also need work to be done to the **required quality standards**. Work quality depends on the knowledge, skill and behavior of the maintenance staff. Knowledge can be enhanced by providing good work procedures, documentation, training and instruction. Similarly, training and experience can help develop skills. The behavior of people is affected by their motivation and morale. Knowing that your supervisor provides feedback, whether favorable or unfavorable, supports you in your learning process, and accepts responsibility for your actions can be very motivating and do wonders for your morale.

We can do a lot to **improve productivity**, not by beating the drum faster, but by removing impediments to work progress. Good logistics support is essential. Thus, getting spares, tools drawings, documentation and procedures on time will make the technician do the job better and faster. Arranging the work day to maximize the time available for hands-on-tools is also very important. Apart from the direct cost savings and uptime improvements, these steps can improve morale significantly.

Plant shutdowns are major maintenance investments, in terms of downtime and costs. The maintenance manager faces a number of risks in organizing and executing them. The major challenges are in managing safety, scope changes and costs.

In order to manage maintenance effectively, we need good information. This is distilled from raw data entered in the maintenance management database. Whether we use fixed format or free text reporting, we can still get good information, provided that the data inputs are of acceptable quality. This is an area where we need to convince the technicians of the importance of good data.

We would like to close by reminding readers that maintenance is an investment that adds value, by ensuring the integrity and profitability of the plant. All investments need a return; it is our job to compute and demonstrate the value we add by doing maintenance.

Appendix 12-1

MAINTENANCE IN A BROADER CONTEXT

In Appendix 9-1, we looked at maintenance holistically, extending the definition to include *the health of the population, law and order, or the reputation of a business*. We noted that we have to manage risks in these cases as well, so we can apply the event escalation and damage limitation models. In this section, we will continue the earlier discussion with some additional examples.

12-1.1 Public health

In an article entitled 'Plagued by cures' in *The Economist*[1], the author argues that preventing diseases in infancy may be a mixed blessing. The study of hospital admissions (for severe cases of malaria in Kenya and Gambia) showed some unexpected results. The admission rate for children with severe malaria was low in areas where transmission of the disease was highest, and high in areas where its transmission was more modest. In these cases, widespread prevalence of a mild form of malaria appears to influence the onset of the virulent form of the disease.

In the same article, the author quotes a study in Guinea-Bissau, where children who had measles were less prone to allergies causing illnesses such as asthma or hay-fever when they grow older. Several other studies show similar results with other childhood infections. Thus, in a more general sense, childhood diseases seem to reduce proneness to other diseases later in life. The human body's immune system is the event escalation barrier, and childhood infections appear to influence its availability.

Historically, we used vaccines to prevent the onset of disease in healthy people and therapeutic drugs to cure sick people. Now, a new generation of vaccines is becoming available to cure the sick, thereby acquiring a therapeutic role. These vaccines are a result of advances in biotechnology and prompt the immune system to cure diseases such as hepatitis-B and herpes. Some 75 new vaccines of this kind are under development, according to an article entitled "Big Shots" in *The Economist*[2]. The body's immune system is still the barrier that prevents the escalation, but such vaccines may be able to increase the barrier availability.

A very successful anti-AIDS campaign is being conducted in Senegal. This commenced in 1986, before the disease had spread in the country. In spite of its predominantly Islamic and Catholic population, the responsible agencies were able to provide sex education in schools. They sold condoms at heavily discounted prices. They targeted the Army, as it had a large group of young sexually active men. In an article entitled "An Ounce of Prevention," *The Economist*[3], reports that a recent survey shows that Senegal indeed appears to have succeeded in controlling the spread of AIDS.

Diet is another area of interest when dealing with public health. Trace amounts of zinc in the diets of children are proving to be successful in reducing the incidence of a wide range of diseases. These include malaria, bacterial pneumonia, and diarrhea. Zinc administered to pregnant women raises the level of antibodies in the blood of their offspring, indicating a better immune system. These children had a lower probability of falling ill in their first-year. In an article entitled "Lost Without a Trace," in *The Economist*[4], the author notes that zinc supplements may soon join iron and folic acid as routine supplements for pregnant women. By improving the immune system, zinc appears to increase the availability of the human body's internal barrier.

12-1.2 Law and order

In Appendix 9-1, we discussed crowd control situations, especially in the context of football hooligans. The use of the *people, plant,* and *procedures* barriers clearly assists the police in maintaining law and order in these situations.

In the United Kingdom, the police support and encourage Neighborhood Watch schemes. People living in a locality form a loose association to protect their neighborhood from vandals and criminals. The police assist them by providing some basic instructions and training. The scheme coordinator keeps in touch with the members and the police. The members assist one another in preventing untoward incidents by remaining vigilant. They also try to improve road safety in their locality, especially if small children are at risk. Such schemes can reduce petty crime and vandalism, and act as a barrier against escalation to more serious offenses.

In an article about the falling crime rates all over the United States, *The Economist*[5] discusses possible reasons, amongst them one of zero-tolerance. An earlier article entitled "Broken Windows" in the *Atlantic Monthly* of March 1982 argues for such a policy. Minor infractions of the law—dropping litter on roads or painting graffiti on walls—become punishable offenses. This produces a climate where serious crimes are unable to flourish. A few years ago, the New York City police commissioner moved police officers away from desk jobs and back on the beat. They were also better armed and given a greater latitude in decision-making. Precinct commanders were held accountable for reducing crimes, not for speed of response to calls, as was the earlier practice. The better visibility and improved morale of the police proved successful in reducing crime rates dramatically.

The Boston Police Department has run a very successful campaign against juvenile crime. Officers and civilians cooperate in scrubbing off graffiti and run youth clubs.They provide counseling services and look out for truants. Juvenile crime rates have fallen dramatically and it is reasonable to link these results to the efforts of the police.

Referring to our risk limitation model in Chapter 9 (refer to Figure 9.1), we note that some of the above steps reduce the demand rate, others act as barriers to prevent event escalation. For example, greater police visibility means that people know that their response is quicker. This stops potential criminals even before they start, thus reducing the demand rate. If a holdup or other crime is already under way, the speedy arrival of the police can prevent further escalation.

12-1.3 Reputation management

In Appendix 9-1, we discussed two cases, one relating to clothing manufacturer Levi Strauss and the other to pharmaceutical manufacturer Johnson & Johnson. Both organizations had built up good reputations over the years with their customers and staff. When they faced very difficult situations, they enjoyed the full support of their customers and staff.

In our risk model, we can think of minor customer complaints as process deviations. If the organization has trained staff and a proper complaint-handling procedure, these minor issues will not escalate into significant grievances. In Section 12.5, we discussed the benefits of keeping open lines of communications with all the stakeholders. Without the benefit of a sympathetic public, a large organization is one more Goliath and by implication, an oppressor. Hence, it becomes even more important for them to build trust with their stakeholders. By doing so, they improve the damage-limitation barrier availability, and this can protect the organization from serious loss of reputation.

12-1.4 Natural disasters

Every year, in the summer months, forest or bush fires rage in many parts of the world. They cause economic and environmental damage as well as casualties among wildlife and human populations. Conventional fire fighting methods are often ineffective. Strong winds, which can accompany these fires, make it very difficult to control their rapid spread. Fire fighters build artificial barriers to prevent the spread by denuding wide swaths of vegetation across the path of the fire. They often do this by burning the vegetation using well-controlled fires. When the main fire reaches this band of burnt out vegetation, it is unable to jump across this artificial barrier.

Storms, typhoons, and tornadoes strike some parts of the world quite regularly. Their energy levels are such that they can cause massive destruction. Usually the most effective solution is to evacuate the population, using early warning methods. By relocating the potential victims, we reduce the number of people at risk.

References

1) "Plagued by Cures." 1997. *The Economist*. November 22: 135-136.
2) "Big Shots." 1998. *The Economist*. May 9: 93.
3) "An Ounce of Prevention...." 1998. *The Economist*. July 4: 109.
4) "Lost Without a Trace." 1998. *The Economist*. August 1: 80-81.
5) "Defeating the Bad Guys." 1998. *The Economist*. October 3: 79-80.

Glossary

The following is a list of terms used, along with their meaning or definition as applied in this book.

Accelerated test	A test in which the applied stress is higher than design values so as to reduce the time to failure. The basic failure mechanism and failure modes must not be altered in this process of acceleration.
Age-exploration	A method used to decide maintenance intervals when failure rates are unavailable. We choose an initial interval based on experience, engineering judgement or vendor recommendations. Thereafter we refine the intervals based on the condition of the equipment when inspected. Each new inspection record adds to this knowledge, and using these we make further adjustments to the maintenance intervals.
Availability	1) The ability of an item to perform its function under given conditions. 2) The proportion of a given time interval that an item or system is able to fulfil its function. Availability = {timc in operation - (planned + unplanned) downtime} / time in operation
Breakdown	Failure resulting in an immediate loss of product or impairment of technical integrity.
Circadian rhythm	A natural biological cycle lasting approximately 24 hours, which governs sleep and waking patterns.
Compliance	A measurement of the percentage completion at the end of a defined period of the routine maintenance jobs due in that period.
Condition Based Maintenance	The preventive maintenance initiated as a result of knowledge of the condition of an item from routine or continuous monitoring
Condition Monitoring	The continuous or periodic measurement and interpretation of data to indicate the condition of an item to determine the need for maintenance
Conformance	Proof that a product or service has met the specified requirements.
Corrective Maintenance	1) The maintenance carried out after a failure has occurred and intended to restore an item to a state in which it can perform its required function 2) Any non-routine work other than breakdown work required to bring equipment back to a fit for purpose standard and arising from: - defects found during the execution of routine work - defects found as a result of inspection, condition monitoring, observation or any other activity.
Defect	An adverse deviation from the specified condition of an item.

Diagnosis	The art or act of deciding from symptoms the nature of a fault.
Disruptive stress	The physical or mental stress a person feels that threatens, frightens, angers or worries a person, resulting in poor or ineffective performance.
Down Time	The period of time during which an item is not in a condition to perform its intended function.
Efficiency	The percentage of total system production potential actually achieved compared to the potential full output of the system.
End-to end testing	A test in which the sensor, control unit and executive element of a control loop are all called into action.
Ergonomics	The science that matches human capabilities, limitations and needs with that of the work environment.
Evident failure	A failure that on its own can be recognized by an operator in the normal course of duty.
Facilitative stress	The physical or mental stress that stimulates a person to work at optimum performance levels.
Fail safe	A design property of an item that prevents its failures being critical to the system.
Failure	The termination of the ability of an item to perform any or all of its functions
Failure cause	The initiator of the process by which deterioration begins, resulting ultimately in failure.
Failure effect	The consequence of a failure mode on the function or status of an item.
Failure mode	The effect by which we recognize a failure.
Failure Modes and Effects Analysis	A structured qualitative method involving the identification of the functions, functional failures and failure modes of a system, and the local and wider effects of such failures.
Fatigue	The reduction in resistance to failure as a result of repeated or cyclical application of stresses on an item.
Fault	An unexpected deviation from requirements which would require considered action regarding the degree of acceptability
Function	The role or purpose for which an item exists. This is usually stated as a set of requirements with specified performance standards.
Hidden failure	A failure that on its own cannot be recognized by an operator in the normal course of duty. A second event or failure is required to identify a hidden failure.
Incipiency	Progressive performance deterioration which can be measured using instruments.
Inspection	Those activities carried out to determine whether an asset is maintaining its required level of functionality and integrity and the rate of change (if any) in these levels.
Instrument protective systems	These instruments protect equipment from high consequence failures by tripping them when pre-set limits are exceeded.
Item	A system, sub-system, equipment or its component part that can be individually considered, tested or examined.

Life Cycle Costs The total cost of ownership of an item of equipment, taking into account the costs of acquisition, personnel training, operation, maintenance, modification and disposal. It is used to decide between alternative options on offer.

Maintainability The ability of an item, under stated conditions of use, to be retained in or restored to a state in which it can perform its required functions, when maintenance is performed under stated conditions and using prescribed procedures and resources. It is usually characterized by the time required to locate, diagnose and rectify a fault.

Maintenance The combination of all technical and associated administrative actions intended to retain an item in or restore it to a state in which it can perform its required function

Maintenance Strategy Framework of actions to prevent or mitigate the consequences of failure in order to meet business objectives.The strategy may be defined at a number of levels (i.e.corporate, system, equipment, or failure modes).

Mean availability With non-repairable items, the point availability has the same value as the survival probability or reliability. As this varies over time, the average value of the point availability is the mean availability.

Modification An alteration made to a physical item or software, usually resulting in an improvement in performance and usually carried out as the result of a design change

Net Positive Suction Head The difference between the suction pressure of a pump and the vapor pressure of the fluid, measured at the impeller inlet.

Non Routine Maintenance Any maintenance work which is not undertaken on a periodic time basis.

Operational Integrity The continuing ability of a facility to produce as designed and forecast.

Outage The state of an item being unable to perform its required function

Overhaul A comprehensive examination and restoration of an item, or a major part of it, to an acceptable condition

Partial closure tests When total closure of executive elements is technically or economically undesirable, the movement of the executive element is physically restrained. Such tests prove that these elements would have closed in a real emergency.

Performance Indicator A variable, derived from one or more measurable parameters, which when compared with a target level or trend, provides an indication of the degree of control being exercised over a process (e.g work efficiency, equipment availability).

Planned Maintenance The maintenance organized and carried out with forethought, control and the use of records, to a pre-determined plan..

Population stereotype The behavior expected of people or equipment (e.g.valves are expected to close when the wheel is turned clockwise). Under severe stress or trauma, people do not behave as trained or according to procedure, they revert to a population stereotype.

Preventive Maintenance The maintenance carried out at pre-determined intervals or corresponding to prescribed criteria and intended to reduce the probability of failure or the performance degradation of an item

Redundancy	The spare capacity which exists in a given system which enables it to tolerate failure of individual equipment items without total loss of function over an extended period of time.
Reliability	The probability that an item or system will fulfil its function when required under given conditions.
Reliability Centered Maintenance	A structured and auditable method for establishing the appropriate maintenance strategies for an asset in its operating context.
Reliability Characteristics	Quantities used to express reliability in numerical terms
Repair	To restore an item to an acceptable condition by the adjustment, renewal, replacement or mending of misaligned, worn, damaged or corroded parts.
Resources	Inputs necessary to carry out an activity (i.e.people, money, tools, materials, equipment).
Risk	The combined effect of the probability of occurrence of an undesirable event and the magnitude of the event.
Routine Maintenance	Maintenance work of a repetitive nature which is undertaken on a periodic time (or equivalent) basis.
Safety	Freedom from conditions that can cause death, injury, occupational illness or damage to asset value or the environment.
Shutdown	A term designating a complete stoppage of production in a plant, system or sub-system to enable planned or unplanned maintenance work to be carried out. Planned shutdowns are usually periods of significant inspection and maintenance activity, carried out periodically.
Shutdown Maintenance	Maintenance which can only be carried out when the item is out of service.
Standby Time	The time for which an item or system is available if required, but not used.
System Effectiveness	The probability that a system will meet its operational demand within a given time under specified operating conditions. It is a characteristic of the design, and may be evaluated by comparing the actual volumetric flow to that theoretically possible when there are no restrictions at the input or output ends of the system.
Technical Integrity	Absence, during specified operation of a facility, of foreseeable risk of failure endangering safety of personnel, environment or asset value.
Test interval	The elapsed time between the initiation of identical tests on an item to evaluate its state or condition. Inverse of test frequency.
Turnaround	A term used in North America meaning planned shutdown. See Shutdown above.
Work Order	Work which has been approved for scheduling and execution.Materials, tools and equipment can then be ordered and labor availability determined.

Index